The Amazing Growth and Journey of UAVs & Ballistic Missiles Defence Capabilities

Where the Technology is leading to

The Amazing Growth and Journey of UAVs & Ballistic Missiles Defence Capabilities
Where the Technology is leading to

by

Lt Gen Vijay Kumar Saxena, VSM

(Established 1870)

United Service Institute of India
New Delhi

Vij Books India Pvt Ltd
New Delhi (India)

Published by

Vij Books India Pvt Ltd
(Publishers, Distributors & Importers)
2/19, Ansari Road
Delhi – 110 002
Phones: 91-11-43596460, 91-11-47340674
Fax: 91-11-47340674
e-mail: vijbooks@rediffmail.com

Copyright © 2013, United Service Institution of India, New Delhi

ISBN: 978-93-82652-13-7

All rights reserved.

No part of this book may be reproduced, stored in a retrieval system, transmitted or utilized in any form or by any means, electronic, mechanical, photocopying, recording or otherwise, without the prior permission of the copyright owner. Application for such permission should be addressed to the publisher.

Disclaimer

The views expressed in this book are those of the author in his personal capacity. These do not have any official endorsement.

CONTENTS

Chapter-1

The Amazing Growth And Journey Of Unmanned Aerial Vehicles And Where The Technology Is Leading To	1
Bringing Down The UAVs : A Growing Imbalance	26
Conclusion and Some Take Aways	35

Figures:

Fig 1 : Vintage Air Threat	3
Fig 2 : F-22 Raptor, Contemporary Air Threat Air	3
Fig 3 : Use of Balloons as Aerial Threat Vehicle	4
Fig 4 : Kettering Bug	5
Fig 5 : Reginald Denny with 'RP-1'	5
Fig 6 : Mid 1930 - ' Queen Bee' Drone	5
Fig 7 : ADM-20 Quail	6
Fig 8 : Ryan Firebee	6
Fig 9 : AQM- 34	6
Fig 10 : Ryan 147 Lightning Bug Series	6
Fig 11 : Teledyne Ryan 1124	7
Fig 12 : Scout	7
Fig 13 : Pioneer UAV	8
Fig 14 : Pointer UAV	8
Fig 15 : Predator	8
Fig 16 : Luna	8
Fig 17 : HALE UAV 'Global Hawk'	9
Fig 18 : Predator Firing Hellfire Missile	12
Fig 19 : Comanche	15

Fig 20 : X 47-B	16
Fig 21 : F-35 Lightening II	16
Fig 22 : Dassaults Aviation nEUROn	17
Fig 23 : Boeing X-45	17
Fig 24 : Northrop Grumans' Firescout Unmanned Autonomous Helicopter	18
Fig 25 : Morphing Technology	20
Fig 26 : Amazing World of Orinthropters	21
Fig 27 : Humming Bird	22
Fig 28 : Black Widow	22
Fig 29 : Desert Hawk	22
Fig 30 : Switchblade Suicide Drone	22
Fig 31 : SWARMs of Quadrotors	23
Fig 32 : Artists Impression of Employment of Boeings' SWARMs	24
Fig 33 : Artists Impression of SWAVNET	25
Fig 34 : RQ 170 Sentinel	25
Fig 35 : Phantom Ray UAV	25
Fig 36 : nEUROn	25
Fig 37 : Phased Array Radar Technology	27
Fig 38 : OSA-AK	29
Fig 39 : Star Streak	29
Fig 40 : Strela 10M	29
Fig 41 : Spyder	30
Fig 42 : Barak	30
Fig 43 : US RQ 170 Sentinel Brought Down by Iran	31
Fig 44 : Laser Avenger System	33
Fig 45 : LASER CIWS Shoots down a UAV	33
Fig 46 : Cougar Anti UAV System	33
Fig 47 : Peregrine Eagle : Anti UAV System	34

Chapter - 2

The Amazing Growth And Journey Of Ballistic Missile Capabilities   37

Contents

Where The Technology Is Leading To	51
Trends	56
The Global Look-See	56
Synergizing Sensor Capabilities in Multiple Domains	60
Optimising Battle Management	70

Figures:

Fig 1 : 3000 + German V2 Rockets Targeted the City of London and Antwerp During WW II	39
Fig 2: SCUD	40
Fig 3: DF-21	40
Fig 4: SHAHEEN III	40
Fig 5: PEACEKEEPER (USSR) SRBM (CHINA) MRBM (PAK) IRBM (US) ICBM	40
Fig 6 : For Ballistic Missiles, There is No Target Too Far	41
Fig 7 : The Eternal Cause Effect Relationship	41
Fig 8 : Spartan BMD Missile (US)	42
Fig 9 : Gazelle BMD Missile (USSR)	42
Fig 10 : The HOE	43
Fig 11 : The HOE Interceptor in the End Game	43
Fig 12 : The SDI - The Big Dreams	44
Fig 13 : Patriot - Lower Tier System	46
Fig 14 : MEADS - Lower Tier System	47
Fig 15 : Aegis - Lower Tier System	48
Fig 16 : Arrow (Hertz) - Ballistic Missile System	48
Fig 17 : S-300 PMU-1	49
Fig 18 : S-300 PMU-2	49
Fig 19 : S-300 PMU-V	49
Fig 20 : S-400 (TRIUMF)	49
Fig 21 : THAAD - Upper Tier System	50
Fig 22 : Pillbox Radar/ Russian A-135 System	51
Fig 23 : US Ground Based and Mid Course Defence System	51

Fig 24 : Ever- Moving Wheels of Time	52
Fig 25 :Turning Face of Technology	52
Fig 26 : Challenges of BPI	52
Fig 27-: Virtues and Challenges of Boost Phase Interception	54
Fig 28 : Complexities of Post-Boost Interception	55
Fig 29 : Challenges of BMD at Re-entry	55
Fig 30 : Satellite Surveillance	57
Fig 31: US SBIRS System Complex	58
Fig 32 : Chinese Satellite Constellation	59
Fig 33 : Russian Satellites in GEO/LEO/HEO	59
Fig 34: Cobra Dane Radar	61
Fig 35 : Mobile Sea Based X Band Radar	61
Fig 36 : Chinese OTH-B Radar Coverage	61
Fig 37 : Voronezh-M Radar Station	62
Fig 38 : Dnepr Pulsed Radar Site	62
Fig 39 : Daryal Radar Site	62
Fig 40 : Dnestr Radar Complex	62
Fig 41 : Nuances of Range-Time Deficit during Interception	63
Fig 42 : Intense Relativistic Electron Beam	64
Fig 43 : Depiction of DE Weapons in Action	65
Fig 44 : Delivering the Kill Energy	66
Fig 45 : Electronic Kill using HPM or CPB Weapons	66
Fig 46: A Futuristic Thermal Kill Weapon	67
Fig 47 : The Eternal Cause Effect Battle	68
Fig 48 : Challenges of IR Homing	68
Fig 49 : The Airborne Laser Test Bed	70
Fig 50 : System of Systems Approach	72
Fig 51 : Joint Tactical Ground Station (JTAGS)	72
Fig 52 : Chinese ASAT Test	73
Index	77

THE AMAZING GROWTH AND JOURNEY OF UNMANNED AERIAL VEHICLES AND WHERE THE TECHNOLOGY IS LEADING TO

THE AMAZING GROWTH AND JOURNEY OF UNMANNED AERIAL VEHICLES AND WHERE THE TECHNOLOGY IS LEADING TO

'When you're out in your backyard this summer, smile - you might be on camera'

- 'The Rise of Unmanned Aerial Vehicles'[1]

Anchor Thoughts. This Chapter is based on two anchor thoughts. Firstly, it aims to convey that UAVs and UCAVs are the aerial threat vehicles whose time has come and secondly, it dwells on the current state of capability in bringing down a UAV and what the technology has to offer in this field today.

The Threat Metamorphosis. Starting from its humble beginning, when the air threat was prosecuted by the valiant air warriors standing up against all odds in attacking aircrafts and mechanically directing munitions on to their targets through a manual hit and trial procedure, **today is an era of multiple air threat vehicles**. These vehicles are technologically enabled for precision, long range, deep strike with stand-off capability and enjoy a high degree of survivability through EW muscle, stealth and manoeuvre. Their multiple-dimension arsenal comprises of many a smart/ intelligent/ lethal munitions with sub-meter accuracy.

Fig 1 : Vintage Air Threat

Fig 2 : F-22 Raptor, Contemporary Air Threat Air

---

1 Kathryn A .Wolfe. Politico, 'The Rise of Unmanned Aerial Vehicles'; www.politico.com/news/stories/ 0312/73901.html. Accessed on 03 December 12.

The UAV/UCAV Evolutionary Path. Alongwith the evolution of combat air power over time, the UAVs and UCAVs have also charted a fascinating evolutionary path. Making a beginning in the mid Nineteenth Century, when on 22 Aug 1849, the Austrians attacked the Italian city of Venice with unmanned balloons loaded with explosives;[2] the UAVs and UCAVs today are standing shoulder-to-shoulder to manned combat aircraft and gaining in strength each day.

Fig 3 : Use of Balloons as Aerial Threat Vehicle

Against Much Resistance. The historic evolution of UAVs has been marked by a series of inconsistent periods of technological development followed by stagnation and long periods of dormancy. The tough resistance to the 'unmanned' vehicles was mainly from the pre-pilot lobby who maintained the position of '**Air Force for the Pilots**' also famously called the 'White Scarf Syndrome'[3]. Anything that had something to do with 'aircrafts without pilots' was simply despised, until a more 'tolerable' terminology, Remotely 'Piloted' Vehicles (RPVs) got into use. The term RPV had a wee bit of acceptability in the pilot community since the same retained the primacy/presence of the 'pilot', albeit at Ground Control Station.

Baby Steps. UAVs/ RPVs/Drones made a visible presence during and shortly after World War I[4]. Some famous names come to mind; AM Low's Aerial Targets based on radio controlled techniques, Hewitt Sperrys' Automatic Airplane, also Known as 'flying bomb'

---

2 'History of Unmanned Aerial Vehicles', http://en.wikipedia.org/history. Accessed on 11 Dec 12.

3 Carl Builder, 'The Icarus Syndrome : The Role of Air Power', 'New Brunswick, NJ : Transaction Publishers, 1994. Accessed on 14 Dec 12.

4 http://www.the nation.com/article/166124/brief-history-of-drones. Accessed on 14 Dec12.

and the Kettering Bug, the Aerial Torpedo.

Fig 4 : Kettering Bug

Inter-War Period. During the inter war period, the development of radio controlled target drones continued unabated. DH 82B Queen Bee drone and several others made news in 1930. During World II RPVs/ drones appeared in large numbers. The Radio Plane Company of Reginald Denny in UK (where the lady Norma Jeane, later to become the famous Marilyn Monroe was spotted and photographed at her job) made about 15,000 drones for the Army during the War. The period 1928-30 also saw some UAVs with (IR) TV based guidance through optical control. However, up to around this time, the concept of assault drones remained largely unproven[5].

Fig 5 : Reginald Denny with 'RP-1'   Fig 6 : Mid 1930 - ' Queen Bee' Drone

Cold War Years. The cold war years saw the piston powered target drones being used as Basic Training Targets for air defence gunners. The first reported use of drones as decoys was the Mc Donnell Douglas ADM-20 Quail, which was carried by Boeing B-52 Stratofortress bombers to help them penetrate the defended air space. In an another first, B-17 Flying Fortresses were transformed into drones to collect samples from inside the radioactive cloud wherein, they were flown directly above the explosion. Around the

---
5 Taylor, AJP, 'James Book of Remotely Piloted Vehicles'. Accessed on 14 Dec 12.

mid 20th Century and thereabout, reconnaissance/spy drones of various types, ranges and capabilities came to be used in a big way.

Fig 7 : ADM-20 Quail

Vietnam War. During the Vietnam War, a large number of RPVs were used by Americans in a large variety of missions. AQM 34, Lightening Bugs, Compass Copes, Ryan Firebee etc undertook a huge 3435 missions (from Aug 1964 to Apr 1975) spanning a wide variety of roles ranging from photo reconnaissance, jamming SA-2 SAM sites, leaflet dropping (Litterbug), ELINT/SIGINT missions and creating chaff corridors to confuse enemy radars. The entire vertical span from 500 ft to about 60,000 ft, was spanned by various RPVs.

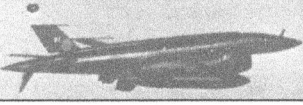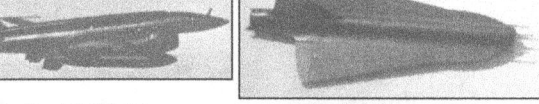

Fig 9 : AQM- 34   Fig 10 : Ryan 147 Lightning Bug Series

Fig 8 : Ryan Firebee

The Arab-Israeli Wars. During the Yom Kippur War the Israelis used Teledyne Ryan 124 R RPVs along with the home-grown Scout and Mastif UAVs for reconnaissance, surveillance and as decoys to draw fire from Arab SAMs. This resulted in Arab forces expending costly and scarce missiles on inappropriate targets[6]. In the Bekka

---

6 http://www.gloria-centre.org>MERIA>Israel; 'Unmanned Aerial Vehicles in the Service of the Israeli'. Accessed on 15 Dec 12.

Valley as well, the Israelis made extensive use of RPVs. As per a report, the Israelis put mini RPVs 'on station' over Syrian airfields using TV cameras to monitor runway activity and transmitting real time information to affect successful interceptions. Israelis also employed other types of unmanned aircraft like Samson and Delilah as decoys. The RPVs with enhanced radar cross section (RCS) confused the Syrian air defence systems which mistook these RPVs as Israeli strike aircraft launched most of their SAMs on them[7].

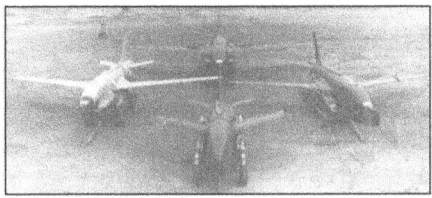

Fig 11 : Teledyne Ryan 1124

Fig 12 : Scout

Desert Storm. During Operation Desert Storm, the US used a variety of UAVs (Israeli Pioneer, Pointer, Midge, Mar, Exodrones etc) for a large number of tasks which included pre-ingress route reconnaissance, chemical agent detection, battle damage assessment, SIGINT/COMINT missions, reconnaissance, surveillance and Command, Control, Communications, Computers, Information & Intelligence (C4I2) missions, directing naval gun fire etc. During this war, one of the most unusual surrenders took place where a Pioneer RPV came over the battlefield, surveying potential targets and five Iraqi Soldiers waved flags at its tiny television camera. It was probably the first time in history when **men surrendered to a robot**[8].

---

7 http://www.airforce.magzine.com/magzinearchive/pages/2002/Jun2002/ 0602.bekaa. aspx, 'The Bekaa Valley War'. Accessed on 15 Dec 12.

8 http:// www.pbs.org/wgbh/pages/fronttime/gulf/weapons/drones/htm. Accessed on 15 Dec 12.

Fig 13 : Pioneer UAV   Fig 14 : Pointer UAV

Bosnia, Kosvo. NATO forces in the Bosnian conflict used the Tier II Predator UAV with one feet resolution Synthetic Aperture Radar (SAR) to monitor the enforcement of ceasefire. Specific tasks included detection of movement of ammunition at night and detection of tampering of mass graves by Bosnian Serbs at night[9]. During the Kosovo conflict, Kosovo Force (KFOR) made extensive use of German multirole UAV LUNA, for real time video dissemination of target information. Based on the reconnaissance by the UAVs, strike aircrafts were launched to neutralize the detected targets. Besides support of air war, UAVs were also used in border surveillance, monitoring ceasefire compliance, combat smuggling and illegal border crossing etc[10].

Fig 15 : Predator   Fig 16 : Luna

Enduring Freedom. During Operation Enduring Freedom, the allied forces made extensive use of UAVs. Following the successful integration of Hellfire missiles on armed version of Predator UAVs, these potent machines saw action in this operation for the first time. In addition, High Altitude Long Range (HALE) RQ4A Global Hawk

---

9 http://www.ukessays.com>Dissertation>Military; 'Employment of UAV in Sub Conventional War'. Accessed on 15 Dec 12.

10 http://www.nato.int/KFOR/docu/pr/2003/10/07.htm/use-of-unmanned-air-vehicle-by-KFOR. Accessed on 18 Dec 12.

UAV was also used extensively. The inputs from Global Hawk and Predator were linked with RC-135, U2, E-8 and other sensor platforms. This enabled combat surveillance of enemy activity and contributed significantly in shortening the kill chain. The UAVs remained on station for long durations, enabling instantaneous attacks upon locating a target, matching it with a weapon and observing the resulting effect. Predator UAVs teamed up with helicopter gunships to provide highly accurate weapon delivery on the ridge lines. This probably marked the first operational employment of a manned and unmanned platform together in the same mission in a synergetic mode. Large scale move of Al-Qaeda and Taliban fighters towards Shah-i-Kot mountains and their assembling and re-equipping was picked up by Predator UAVs resulting in the launch of Operation Anaconda.

Fig 17 : HALE UAV 'Global Hawk'

Iraq/ Afghanistan. During Iraq, as also, in Afghanistan, multiple in and out-of-theatre Ground Control Station (GCS) were used to simultaneously control a number of Predator orbits over battle space providing great flexibility to ground commanders. These UAVs were operated with 'pilot-in-the-loop' concept where US Air force UAV pilots sitting in California controlled aircrafts and mission payloads via satellite data link and distributed imaging and intelligence products on Global Intelligence networks[11][12].

Iraqi Freedom. During Operation Iraqi Freedom, the allied force used at least eleven types of UAVs, Predator, Global Hawk, Reaper, Shadow 200, Hunter, Pointer etc, to name a few. Finder UAV,

---

11 http://www.defence-update.com/products/p/predator.htm. Accessed on 19 Dec 12.

12 http:// www.rand.org>Reports and Bookstore>Research Briefs>RB 9148>operation-enduringfreedom-an assessment/RAND. Accessed on 19 Dec 12.

capable of carrying out chemical and biological sensors was used for sniffing the air over Iraq. Hellfire equipped Predators were very successfully employed to take on chosen targets with precision or to hunt down terrorists in hunter-killer operations. According to statistics, the Global Hawk in ISTAR mission located at least 13 Surface to Air (SAM) Batteries, 50 SAM launchers, 300 canisters and 70 missile transporters. It also imaged about 300 tanks; nearly 38% of Iraqi armed forces[13] [14].

Finding the Needle in the Haystack. The operational use of weaponised UAVs could either be in the direct support of military operations as discussed above, or in **hunter killer missions** (also referred to as search and destroy missions) implying searching and hunting down terrorists from their hideouts. While the former operation normally unfolds in recognised military areas, the latter trespasses civilian areas with huge collateral damages.

GWOT. The Global War on Terror (GWOT) in Afghanistan-Pakistan region, has witnessed a huge increase in the hunter killer missions using armed drones like Predator etc. As per statistics, the drone strikes have steadily increased from two in 2006 to three in 2007 to 34 in 2008 to 41 in 2009 and counting. According to another report in the open domain, as of Oct 2012, US has conducted more than 300 drone strikes killing at least 2500 people, many of them civilians[15]. It has taken 16 strikes, 14 months and an additional 207-321 additional deaths to take out the dreaded terrorist Baitullah Mehsud, the leader of Pakistani Taliban[16].

Collateral- A Huge Negative Issue. The question of collateral damage is a huge negative issue bringing into question the claim to precision and surgicality in the armed UAV/drone operations. Experts feel that in find-fix-finish type of hunter killer missions using armed UAVs, collateral damage creates more militants than

---

13 http://www.iar.gwu.org/node/144/drone-wars-armed-unarmed-aerial-vehicles-international-affairs. Accessed on 19 Dec 12.

14 http://www.en.wikipedia.org/wiki/history-of-unmanned-aerial-vehicles. Accessed on 19 Dec 12.

15 http://www.huffingtonpost.com/.../ afpak-anniversary : 11 years gone, 128500 direct deaths. Accessed on 19 Dec 12.

16 Andrew Callam, 'Drones Wars : Armed Unmanned Aerial Vehicles', International Affairs Review Volume XVIII, No3, Winter 2010. Accessed on 19 Dec 12.

it eliminates. It is from here that the felt need for a manned and unmanned capability in joint missions, arises.

Drones and More Drones. Notwithstanding all the above, the use of drones the world over has seen a dramatic rise. An interesting report published in the Indian Express on 11 Nov 2012 informs us that when President George W Bush declared War on Terror 11 years ago, the Pentagon had fewer than 50 drones. Now it has around 7500. 70 countries own some type of drones or the other, China, in a single air show (Zhuhai Air Show 2010) unveiled 25 new drone models. There are as much as 680 active drone development programmes currently being pursued worldwide with a global market for R&D pegged at $ 11.4 billion. On the flip side, the collateral damage from drone strikes, especially in the GWOT is becoming a huge issue in Pakistan. According to a Study conducted by Stanford Law School and New York University School of Law, an estimated 1908 to 3225 people have been killed in Pakistan in the last eight years while only 2 percent of those killed are high value targets[17].

UAVs/UCAVs- Versatile Threat Vehicles. There are many virtues, which make UAVs a versatile threat vehicle. There are no crew fatalities, no crew fatigue, no training deficit on board, given the expertise of the pilot at the ground control station. As the technology has shown, the weapon loads lifted by UAVs and UCAVs have become comparable to manned aircrafts. Sample the weapon load of Predator (B) - 'Reaper', the largest and the most powerful variant of Predator UAV. It can carry eight laser guided AGM 114 Hellfire Missiles besides two GBU 12 Paveway II Laser Guided Bombs and a multi-spectral targeting system, complete with its infrared sensor, TV Camera, image intensifier night camera, a laser designator and a laser illuminator. Hellfire missile could be exchanged with AIM 9 Sidewinder missile or another PGM[18]. A machine with such load and an endurance of 42 hrs (with 1000 lb) or 14 hrs (when fully loaded) is a formidable platform, indeed comparable to manned aircrafts.

---

17 Indian Express, Bhubaneswar edition 11 Nov 12, Ed, CENTERPIECE. http// www. Indian express.com>story. Accessed on 20 Dec 12.

18 **Reaper : A New Way to Wage a War**, Time: 40.01 Jun 2009. Accessed on 20 Dec 12.

Fig 18 : Predator Firing Hellfire Missile

All Terrain Capability. UAVs also score over manned platforms in terms of having no restrictions to operate in hazardous/ contaminated environment. The range, reach and precision combined with tremendous endurance (Global Hawk UAV holds World Endurance Record of 30 hrs, 24 minutes, 01 seconds at an absolute altitude of 65,381 ft)[19]. In addition to this, as stated, the possible mission spectrum of UAV/UCAVs is indeed expansive– Pre-ingress route reconnaissance NBC detection/ sampling, battle damage assessment, EW, SIGINT, COMINT, ISR/ C4I2 mission, precision attacks on chosen targets of interest, directing naval gun fire etc, to name a few. The strength of these machines lies in their open architecture, flexible mission payloads, video-on-demand capability, instant connect (typical time for a control message to UAV on a satellite downlink is 1-2 seconds), capability for dynamic in-flight re-tasking, high manoeuvre, high strength, high immunity and near real-time high data transfer rates.

A Futile Comparison. In light of the growing clout of UAV and UCAVs, there is much talk of their comparison with manned platforms by counting the virtues of one against the other. This is a futile exercise since the UAVs/ UCAVs are not a complete replacement of the combat aircraft and the two are likely to co-exist in a non-mutually exclusive domain in a net centric battlefield. The idea of co-existence actually stems from the respective vulnerabilities and limitations in both the manned and unmanned platforms and how both can largely address each other's sub-optimalities in joint operations.

---

19 **Aviation History as Global Hawk Completes US-Australia Flight**. Australian MoD Press Release 24 Apr 2001 carried in http//www.en.wikipedia.org/wiki/Northrop Gruman RQ-4-Global Hawk. Accessed on 20 Dec 2012.

The Weak Muscle of UAVs/ UCAVs. While any number of combat virtues may be counted in favour of UAVs/UCAVs, these machines are limited on many counts. Most prominent is their vulnerability to minimal air defence. While the Predator UCAVs succeeded in striking down several Iraqi radar units during operation Iraqi Freedom, these quickly became targets for Iraqi air defence. The Iraqi Air Force shot down three Predators with relative ease. Even in areas where there is no strong air defence, UAVs are vulnerable to ground fire from small arms (SA) weapons/ rockets, normally held by ground forces or even with terrorists. For example, even in the tribal region of Pakistan, where there is virtually no air defence, Taliban has claimed to have shot down several CIA drones over South Waziristan[20]. Recently, Iran's Defence Minister confirmed that their SU-25 Frog foot Warplane had shot down a US Predator drone. This is the first killing of a US UAV in international air space over the Gulf [21]. Even when not facing enemy fire, Predator statistics show that it has crashed due to mechanical error 43 times per 100,000 flying hours, whereas a typical manned aircraft crash data shows 2 per 100,000 flying hours. Lack of military presence on ground also limits the capability of the drones to assist in acquiring critical intelligence. Hunter killer operation can only eliminate terrorists but fail to get intelligence as 'dead men tell no tales'.

Other Limitations. Lack of multiple intelligence agencies also inhibit the drones to identify the targets. First few months after 9/11, a Predator pilot spotted a tall man in flowing white robes walking near the Eastern border of Afghanistan. Incorrectly believing the man to be Osama-Bin-Laden, it fired a missile killing the innocent villager. In fact, co-laterals albeit unintentional, could give a fillip to further acts of terrorism in retaliation. Taliban carried out its Mar 2009 attack on Lahore Police Academy in retaliation to the continued drone attacks[22].

Vulnerability to Jamming. UAVs/ UCAVs driven by the way point navigation system of the GPS are perennially vulnerable to jamming for which jammers are freely available on the internet. The latest

---

20 http://www.reuters.com/02/Pakistan-USA-drone>taliban-militants-say-they-shot down- US-drone. Accessed on 23 Dec 12.

21 'Yes, We Fired at US Drone : Iran': Indian Express Bhubaneswar Edition 08 Nov 12.

22 16 ibid. Accessed on 16 Dec 12.

in this is the 'spoofing' of a GPS receiver on board a UAV, thus taking control of the UAV, much in the same manner as hijacking a plane. The UAV/ UCAV thus taken under control can be made to do what the spoofer wants; viz, to crash on a safe area or to be retrieved safely for study and analysis and worse still, to be used as a missile for suicide mode crashing on desired targets, much like 9/11 mode of attack[23]. Also, a UAV cannot take a decision in the situation of ambiguity, nor can it take on opportunity targets that come up suddenly. It is a sitting duck for strong air defence and is devoid of pilot's 'hunch' and IQ.

The Win-Win. The win-win therefore is to combine the genius of human intelligence, intuition, grit, tolerance for ambiguity, instant decision making capability in a dynamic manner, IQ, hunch and experience of a combat fighter with the brute precision, range, stealth, all terrain/ all weather/ all environment operability, immunity and versatility of the UAV/ UCAVs. This deadly combination will provide an integrated battle winning capability by a force multiplicative effect, duly supported by seamless command, control and battle management systems in an automated and net-centric hierarchy. Besides, this will result in seamlessly threading of the sensor-shooter cycle.

Future Thought. The above is the emerging thought on the future of drone warfare. The same is propelling research in UAV/ UCAVs on one end and integration efforts of manned with unmanned at the other. Basically, the 3D feature of the UAV/ UCAV is being explored, where, the three Ds, stand one each for Dull (implying long endurance), Dirty (Any terrain no restriction) and Dangerous (implying lethality). In consonance with this thought, the US Defence Advanced Research Project Agency (DARPA) is working on one such integration with APACHE Armed Attack Helicopter (AAH) and COMANCHE UAV. Also, keeping in mind that a large number of UAV accidents take place due to human/ UAV interaction delays/ default, the combined missions will reduce this data-link vulnerability, decrease data transit time, provide visual confirmation of UAV actions to the combat pilot and enhance the surveillance and kill capability in the hands of the pilot, multiplicatively.

---

23 http://www.foxnews.com/drones-vulnerable-to-terrorist-hijacking-researches-say. Accessed on 23 Dec 12

Fig 19 : Comanche

Other Ventures. Another Study in this aspect has been undertaken involving A-10, F-16 and AC-130 pilots in a possible joint operation with Predator UAV. The aim is to determine appropriate level of pilot control and UAV/ human interaction. In this Study, interesting facts got debated, like placement combination options for manned/ unmanned aerial vehicles. UAV status display for lead aircrafts, pilot community's unanimous recommendation of never to permit UAVs to perform offensive action on their own, viz, re-striking a target after a miss, pilot-in-the loop requirement, options for letting some UAVs leave the formation, giving control to UAVs to lead manned aircraft in certain situations, suitability of a dual seat/ multiple crew aircraft over single seater in a manned-unmanned combination, limiting the use of UAV as an advanced tool/ weapon with decision authority given to the pilot and not a machine-over-man scenario[24].

Future Research Areas. There are many new areas of future technological research in the field of UAVs. Some of these include miniaturization of electronics, qualitative improvement in sensors and building reliable and jam resistant data links. Continuous research and trials are also taking place in operation of UAVs from aircraft carriers. On 7-10 July 2012, a US Navy experimental UAV X-47B was launched and recovered safely from an aircraft carrier[25].

---

24 http://www.cow.mit.edn/courses/aeronautics-and-the-integration-unmanned-aerial-vehicles- in-offensive- combat-situation. Assessed on 23 Dec 12.

25 http://www.navy.mil/submit/display.asp.story_id66319.navy-closer-to-landing-UAV-on-aircraft carrier. Accessed on 23 Dec 12.

Fig 20 : X 47-B

Another Trend. In an another trend, the leading combat aircraft manufacturers of the world are working towards launching the unmanned versions of their strike fighters together or shortly after the launch of manned version. In this context, there is an open source information regarding M/S Lockheed Martin planning to launch the unmanned version of their latest F-35 Lightening II Stealthy Joint Strike Fighter soon after the launch of its manned version. Besides the many combat virtues of F-35, the unmanned version can assume riskier mission profiles, eg, mounting external pylons, or carrying increased payload (the useful payload can be increased from 5200 to 18000 lbs albeit with reduced stealth)[26]. Such platforms will also have the capability of STOVL (short take off and vertical landing), making them suitable for operation from aircraft carriers.

Fig 21 : F-35 Lightening II

---

26 http://www.news.USon.org/news-analysis/news/unmanned-joint-strike-fighter. Accessed on 23 Dec 12.

nEUROn Arrives. Dassault Aviation is in the final phase of providing nEUROn, a supersonic, two engined, long range combat drone capable of performing attacks with nuclear weapons. nEUROn is significantly larger and more advanced than other well known UAV systems like MQ-1 Predator. It is roughly of the size of a Mirage 2000. It will have capabilities of air-to-ground attack with precision guided munitions and will rely on an advanced stealth airframe that can penetrate undetected. Another feature being contemplated is the ability to control squad flight in automatic mode from all advanced fighters like Rafael, Euro fighter or JAS-39 Grippen platform. nEUROn, as a stealth technology demonstrator has completed its maiden flight on 01 Dec 12. The flights are to continue in France until 2014.[27] [28] [29] In addition, several number of joint unmanned combat platforms of the size and capability of combat fighters are under development by world's leading manufactures. Some names include Boeings X-45, Northrop Gruman X-47 etc.

Fig 22 : Dassaults Aviation nEUROn   Fig 23 : Boeing X-45

VTOL/SVTOL UAVs. As stated, the another emerging concept is the development of VTOL/STOVL UAVs. Northrop Gruman's MQ-8 Fire Scout is one such unmanned autonomous helicopter to provide reconnaissance, situational awareness and precision targeting support for US ground, air and sea forces. It has an endurance of three hours, a range of 100 nautical miles with modular mission

---

27 http://www.killerapps.foreignpolicy.com/.../france-joins-the-stealth-UAV-club. Accessed on 23 Dec 12.

28 http://www.reuters.com/article/.../id VS 39222/dassault-aviation.the nEUROn makes its maiden flight. Accessed on 23 Dec 12.

29 http://www/en.wikipedia.org/iki/dassault-nEuro. Accessed on 23 Dec 12.

payloads to include Hellfire Missiles, Viper Strike Laser Guided Glide Weapons and has pods carrying the advanced precision kill weapon systems. Laser target designation, gimbaled Electro optical/ IR payloads, day/ night cameras, coupled with laser range finders provide all weather operational capability[30].

Fig 24 : Northrop Grumans' Firescout
Unmanned Autonomous Helicopter

Towards Full Vehicle Autonomy. Another emerging concept in UAV is to achieve full vehicle autonomy. In fact, the field of air vehicle autonomy is a recently emerging field. It aims to reduce the dependence of a UAV/UCAV from the Ground Control infrastructure making it a self-contained and autonomous war machine. Such an end state will call for realization of many cutting edge technologies. Firstly, the sensor-fusion technology must combine information from different sources for single-window usage on-board the vehicle. Motion Planning and Trajectory Generation technologies must determine an optimal path for vehicle to go and execute optimal control manoeuvre or to stick to the path defined and the artificial intelligence tools must provide a degree of autonomy and decision making capability based on dynamic parameters available at the mission end. The idea is to teach the machines to be 'smart' and act more like humans. This is being achieved by using technologies related to artificial intelligence, expert system, neural networks, machine learning, natural language processing and vision. Besides this, decentralised decision making algorithms are being proposed to reduce the communication load to the UAV.

---

30 **Autonomous Fire Scout UAV Lands on Ship,** http://www.aviation.week.com/article.aspx?id=xml/asd-08/requirement-for-larger-fire- Scout-VTOL UAV. Accessed on 23 Dec 12.

Integrated Architectures. An open source article presents the results of collaborative effort between MIT and an agency named, Aurora Flight Sciences to develop new integrated architecture that combines search and track functions and solves the challenging multi-vehicle dynamic target resource allocation problem in real time and in the presence of uncertainty. In this, multiple UAVs are used to collaboratively search the environment and keep track of any target found. UAVs persist in the base line search mode and dynamically transit to the track mode either once a new target is detected or when the 'uncertainty' in a target previously detected, exceeds a desired threshold. The Onboard Planning Module (OPM) is a high-level path and task planner. It assigns the vehicle to search for targets (where number and location are still unknown) or track known targets. Whenever a vehicle is searching, the OPM uses a set of probability distributions to guide the vehicle along an optimal search path that maximizes the likelihood of detection of targets. Images captured from the on-board camera are loaded and analysed by Vision Processing Unit. The Co-operative Search and Track architecture allows multiple vehicles to perform co-ordinated search and track in a co-ordinated fashion. This technology is being used in Hummingbird Nano UAV[31].

Exciting New Technologies. Exciting new technologies are entering the field of UAVs/ UCAVs in almost the same time consonance as for combat aircraft. One such technology is morphing, i.e, changing the configuration of the vehicle itself to maximize performance in different flying conditions. Morphing aircrafts are multi-role that change their external shape substantially to adopt to a changed mission environment during flight[32]. Variable sweep wings, landing and take off flaps, retractable landing gear, variable incidence manoeuvre etc, are some established morphing forms. Morphing in UAVs/ UCAVs is becoming equally visible with possible areas being wing/ fuselage/ engine/ tail. All possible change techniques, viz, folding/ hiding/ telescoping/ expanding/ contracting/ coupling/ de-coupling etc, are being applied to bring changes in aspect ratio, wing area, wing twist and more. This provides variable performance

---

31 '**Increasing Autonomy of UAVs**-Dspace@MIT;http://www.dspace.mit.edn/open access-disseminate/1721.1/51874. Accessed on 23 Dec 12

32 Weisshaar T.A (2006) morphing-aircraft-technology-new-shapes-for-aircrafts-design, http://www.rto.nato.int/abstrach.asp. Accessed on 24 Dec 12.

in different modes, like longer loiter time with efficient dash speeds, also high responsiveness for time critical deployments, high agility to attack fast moving and ground targets and long persistence to dominate large operational areas.

Fig 25 : Morphing Technology

Next Generation Morphing. Besides the conventional morphing configurations, the next generation morphing designs include distributed actuators to increase survivability and decrease weight, stretching or sliding skins that meet the changing requirements of symmetric morphed wings and/ or conformal flaps for flight control. An innovative morphing design by Lockheed Martin hides a substantial portion of the wing area during low altitude, transonic dash portion of the mission. Another morphing design by Raytheon is based on telescoping, which is ideal for controlling aerodynamic drag. The leap of technology indicates that morphing technologies are going to find more and more usage in the UAVs/ UCAVs of today and tomorrow.

A New Variation. Orinthropters are another emerging variation of UAVs/UCAVs. These can fly by flapping their wings just like the birds. M/S Aero Vironment of US is a lead player in this field. Starting its supposedly existential journey from 4th Century BC, where in the epic Ramayana, there is a mention of "Pushpak Vimana", the Orinthropter of the Gods, this innovative design has indeed come of age.

Versatility of Orinthropters. Unmanned Orinthropters have many practical applications. These can carry out aerial reconnaissance without alerting the enemy forces. The designers hope to eliminate the motors and gears of the current design by more closely imitating animal flight muscles. Research is accordingly on for developing a re-reciprocative animal muscle for use in such UAVs. As opposed to a jet engine, which produces relatively narrow stream of fast air, an Orinthropter displaces a comparatively large mass of air at slow speed, thus enhancing efficiency in flight. The lift and thrust generated by the flapping wings can be set to zero angle of attack in its upstroke. This allows the Orinthropter to pass easily through the air, thus reducing drag.[33] [34]

Fig 26 : Amazing World of Orinthropters

Nano Impacts UAVs/ UCAVs. Nanotechnology is impacting the UAV/ UCAV development in a big way, Aero Vironments Hummingbird is only 7.5 centimeters/10 grams with a flying speed in excess of 15 miles per hour with precision hover and live video streaming capability. This surveillance prototype of the DARPA made it to the front cover of the Invention Issue of TIME Magazine (November 28) which featured the same as one of the 50 new inventions.[35] A similar Nano UAV from the same firm is called 'Black Widow'. This micro UAV weighs less than 50 grams. It has a flight speed of 200 feet/ second, endurance of 30 minutes and a hover endurance of 8-10 minutes. It has the capability of controlled transition from outdoor to indoor and precision flying in urban settings. There are several other UAVs in the light/very light range. RQ-11 'Raven' of M/S Aero Vironment is a small hand-launched

---

33 http://www.en.wikipedia.org/wiki/orinthropter. Accessed on 25 Dec 12.

34 http://www.ho.secs.vcla.edn/np-contract/uploads/04/Jpl10-2001.pdf. Accessed on 25 Dec 12.

35 http://www.usavision.com/2011/11/21 Accessed on 25 Dec 12.

UAV which is back-packable and weighs only 1.9 Kilograms. It has a range of 10 kilometers and an endurance of 60-90 minutes. The payload is an Electro Optical (EO) camera and an Infrared (IR) camera with a capability of real-time video imagery, especially for the short tactical requirement of 'over the hill/ around the corner[36]'. Lockheed Martin's 'Desert Hawk' UAV is only 3.2 Kilograms while Aero Vironments' 'Switch Blade' is only 2.5 Kilograms. Research is also on in the field of pseudolite (a contraction of the term pseudo-satellite). It is something that is not a satellite but performs the functions commonly in the domain of satellites. These are often small trans-receivers, which are used to create local ground based GPS alternatives[37].

Fig 27 : Humming Bird   Fig 28 : Black Widow   Fig 29 : Desert Hawk   Fig 30 : Switchblade Suicide Drone

The SWARM (Smart Warfighting Array of Reconfigurable Modules). Another manifestation of nano-technology application in the UAVs is the application of UAV swarm technology. In this, autonomous UAVs are made to fly in a swarm without using the ground control station and delivering continuous co-ordinated information to enhance situational awareness. Individual vehicles in the swarm are so programmed as to communicate and co-ordinate actions of each other and one another to conduct autonomous searches. The data collected is aggregated and relayed back in real time.[38] Small scale UAVs functioning as SWARM, rely upon each other as migrating birds. They can collect data precisely, navigate

---

36 http://www.army-technology.conc/projects/rq11-raven. Accessed on 27 Dec 12.

37 en.wikipedia.org/wiki/pseudolite. Accessed on 03 Jan 13.

38 http://www.designers.com/another.asp?section-'autonomous-uav-fly-in-swarm'. Accessed on 03 Jan 13.

through complex and difficult terrain and are continuously kept updated as to their surrounding, status and their inter-se position with respect to the swarm. This is achieved by using some evolutionary algorithm.

Fig 31 : SWARMs of Quadrotors

Group Redundancy. As a result of the above networking algorithms, if out of a large swarm formation, a few UAVs land due to malfunction or have other problems such as engine failure etc, the control system ensures that the other UAVs become aware of this and form new formations that allows the rest of the swarm to collect the data which the malfunctioning UAVs were supposed to be collecting. This allows the swarm to function as a 'live mind' of sorts, making decisions in the autonomous mode.

Aping Animal Instincts. Boeing is one company that is developing swarm based technology. The same is based heavily on the communication systems used by animal swarms. Suitable algorithms are being developed for performing variety of tasks such as searching a designated area, mapping it and generating way points and generating information which can be relayed to the network in near real time.

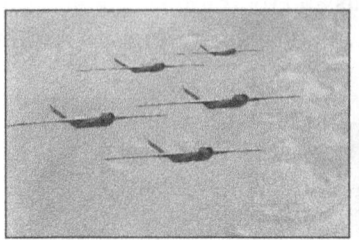 

Fig 32 : Artists Impression of Employment of Boeings' SWARMs

SWARMs - The Future of ISTAR Missions. M/s Ecole Polytechnique Fedarale de Lausanine in Switzerland is developing a 'Swarm Software' (called SMAVNET) for use in disaster situations. SMAVNET or 'Swarming Micro Air Vehicles Network' (each weighing 420 grams and having a 80 centimeter wing span) is being designed as a communication system relay in disaster situations to aid rescuers. The software uses a gyroscope alongwith pressure sensors to make decisions as to which flight path is better than the other. Their communication abilities mimic that of 'ants'. Much like ants, which use pheromones to remember the path they take from their nest to food. The SMAVNET uses similar principle to allow the UAVs to paint a path based on remembrance so that it can have a constant communication path to the home base for sending information back[39]. The day is not far when we will see swarms of operationalised UAVs, each small vehicle lifted and propelled by rotors (four rotors, making the vehicle a quad copter) and moving like a flock, much like the storks or wasps or ants. Each small vehicle will be tied to the other in the soft mode through yet evolving algorithms which will allow them to operate as a 'team'. The future of ISTAR mission is here!

---

39 http://www.blogs.bn.edn/bioaerial 2012/2012/........./swarming-a-team-sport-for-uavs Accessed on 03 Jan 13.

Fig 33 : Artists Impression of SWAVNET

Stealth Impacts UAVs. Stealth technologies are also impacting the modern day UAVs and UCAVs in a very visible manner. Many leading nations in the forefront of UAVs/ UCAVs capabilities, are adopting stealth techniques as a measure to enhance survivability. Lockheed Martin's RQ 170 Sentinel is an example of stealth UAV. This tailless flying wing aircraft has pods, presumably for sensors or SATCOM having been built into the upper surface of each wing. Iran downed one such UAV in December 2011, which was virtually intact. Experts opine that it was probably equipped with cyber/ electronic warfare attack systems. Iran latter claimed to have decoded all data in the captured US stealth drone[40]. It was reported in the open media on November 2012 that Israel was also working on a fairly large UAV with minimal detection features[41]. M/S Boeing's Phantom Ray stealth UAV made its first flight in April 2011. Mention has already been made of 'nEUROn' stealth UAV by M/S Dassault Aviation.

Fig 34 : RQ 170 Sentinel   Fig 35 : Phantom Ray UAV   Fig 36 : nEUROn

---

40 http://www.theaviation.com/category/captured/captured-stealth-drone Accessed on 04 Jan 13.

41 27 ibid. Accessed on 04 Jan 13.

BRINGING DOWN THE UAVS : A GROWING IMBALANCE

A Growing Imbalance. Truly speaking, a well armed and a technologically enabled UAV is as much an attackers' delight as it is a defenders' nightmare. The most prominent talk in the defenders' domain is that of the growing imbalance between the total cost of a UAV/ UCAV vis-à-vis the cost of the SAM/ manned/ ground based kill system that may be deployed to kill it. Since the latter is far-far higher than the former, the major concern is to reduce the cost of kill by using innovative technologies, preferably in the soft-kill mode.

Challenges. Bringing down a UAV/UCAV throws up multiple challenges. The first of all is the detection challenge defined by low radar cross section (RCS), small size, low acoustic signature and low IR signature of the UAV/UCAV. Though Predator/ Reaper type of UAVs/ UCAVs flying at about 100 miles per hour at low altitudes may be a child's play to shoot down for a manned jet fighter, however, if the drone flies higher, it will pose problem for electronic detection due to its lower RCS. For example, the Predator is undetectable beyond 3000 feet and inaudible beyond 1000 feet. The smaller drones have their survivability (implied high degree of difficulty in detection) built into their miniature sizes or stealthy features. Both these features however exploit the advantage of low RCS signature.

Taking On The Challenge. Defenders are busy researching new and innovative methods to detect small UAVs/ UCAVs. The advent of matured phased array radar technology coupled with the tremendous growth in radar computing and processing power have the capability to pick up small maneuverable UAVs using the technology of electronic beam switching.

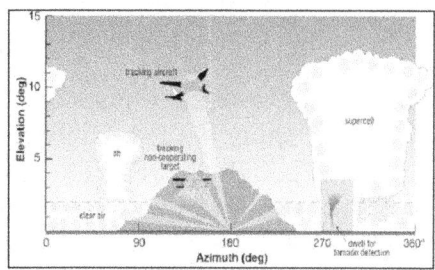

Fig 37 : Phased Array Radar Technology

Optimal Sensors. UAV possesses lower speeds in comparison to aircraft which aids in radar detection. In this scenario pulse doppler radars with MTI facility prove to be quite effective. The optimal frequency bands include S band, X band or a combined S/X band. For detection of small, un-cooperative, low RCS targets in high clutter environment, like the unmanned systems, the trend is to go for higher speeds of scanning incorporating multiple solutions on the same platform, viz, pure RF/ Radar solution, electro-optical solution or even visual warning, as the last resort[42].

Passive Sensing. Another vulnerability of UAVs based on its EM uplink and downlink is being used by passive sensors to perform the function of alerting and cueing. Equipment in this category include IR alerters and ESM acoustic alerters. A typical passive IR detection system based on IR is a two tier system to include a wide-area, panoramic search sensor platform with passive IR sensors and a tracking and verifying sensor platform with a high resolution IR sensor and a Laser Range Finder (LRF). Once a tentative track is established by wide-area search sensors, the verification sensor's line of sight is directed instantaneously to the track co-ordinates indicated by the search sensors with the help of rapidly steerable mirror gimbals. A closed-loop control using the viewer's IR image, links the LRF's line of sight on to the target for precise laser ranging. The target in range velocity is determined by means of computive range measurement. Such silent mode air surveillance systems are normally employed in conjunction with other active systems which aim to have the incoming threat on a closing-in-track with the silent passive system focused on the end game. IR sensor based detection

---

42 http://www.detect-inc.com/security/html/applied-radar-technologies. Accessed on 06 Jan 13.

systems are also ideal for areas in the 'radar shadow' or as a gap-filler in certain geographical areas[43].

ESM Based Sensing. A number of ESM based solutions are also useful in detecting the UAVs through recognition of their RF/ EM signatures. Though the ESM based sensors have been around since World War II, a host of new technologies have entered this field to improve the probability and accuracy of detection. The traditional frequency range is between 1-2 GHz to 18 GHz. Some technologies being put to use include 'Instantaneous Frequency Measurement' and higher speeds in sampling and digital processing of signals[44].

The Acoustic Advantage. Acoustic sensors have the advantage of not being restricted to the line-of-sight operations. Besides, these can have all weather operational capability. They can also deploy search-while-track capability and can provide useful track information about UAVs/ UCAVs. Unlike their electro-magnetic or electro-optic counterparts, acoustic sensors are capable of searching all frequencies and all angles allowing for a wide open range capability. Heat and electro magnetic radiation seeking sensors are easily fooled through countermeasure like dispensing flares or jamming, while to avoid an acoustic sensor, the aerial vehicle must hide its acoustic signature. This is not easy, as many a UAV/ UCAV cannot operate without generating a certain (albeit low) acoustic signature. Acoustic sensors have the further advantage of operation in cloudy or overcast battle environments undeterred by smoke blanketing. On the flip side, such sensors however do not have enormous long range capability. Besides, these are also affected by atmospheric conditions. The presence of mean temperature profiles can cause sound waves to refract, thus reducing the effectiveness of such sensors[45].

The Hard Kill : VSHOARD Domain. Once detected, the UAVs are a soft target for conventional hard kill systems. Conventional fair weather/ all weather VSHORAD systems consists of guns and

---

43 http://www.zmnl.her/tanzekek/etic/konferencia/april 2001/prokob.html/surveillance of - UAVs. Accessed on 16 Jan 13.

44 Ftb.rta.nato.int/public/publicfulltext/RTO/MP/RTO/esm-sensor-for-technical-information-in-air-defence systems. Accessed on 06 Jan 13.

45 http://www.ieeexplore.ieee.or/ie!5/4803720/4806498/04806528.pdf/low-cost-acoustic-array-function-small-uav-detection and tracking. Accessed on 06 Jan 13.

manportable/ short range SAMs, (such as Star Streak, Igla 1M, Strela 10M, OSA-AK, etc) which have the capability to kill a UAV, provided it has been detected, either visually or electronically, by their surveillance /tracking/ missiles guidance radars. During the Kosovo War, machine gun fire from helicopters was effectively used to shoot down UAVs in the visual domain. Yugoslav forces also brought down a few Pioneer UAVs using ground based fire. In the context of UAV kills, the experts, however, count gains vs losses in terms of the number of 'pilots' lives saved' and the 'cost of doing business'[46]. In yet another kill option, UAVs can be pitched against adversary's UAVs in a kinetic-kill mode wherein, a mini/ micro UAV can be used to hit the target UAV. The challenges of initial acquisition, tracking and providing continued guidance to own UAV need to be surmounted. Active seeker/ fire and forget, laser designation, riding on the guidance beam of a mother system and more, are ideas which could be pitched to take on the challenges defined.

Fig 38 : OSA-AK   Fig 39 : Star Streak   Fig 40 : Strela 10M

Hard Kill : SRSAM Domain. Besides the VSHORAD system, designers of several Short Range SAMs (SRSAMs) have claimed that their weapon systems are capable to take on the UAVs/ UCAVs. The Israeli IAI industries claim that their Barak SRSAM, which is designed to be used as a point defence missile system on warships is capable to take on UAVs besides aircraft and anti ship missiles. While Barak-1 has an operational range of 10-12 kilometers, its higher versions (Barak 8) goes up to longer ranges 70/100/120

---

46 http://www.internet data.info/.../uav-decoy-strategies-theories-and-the-modern-art-of-war. Accessed on 06 Jan 13.

kilometers. Similarly, the Spyder (Surface to Air Python and Derby) anti aircraft missile system developed by M/s Rafael of Israel is claimed to be effective against unmanned systems.

 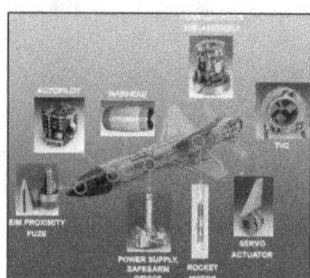

Fig 41 : Spyder   Fig 42 : Barak

Manned Aircrafts. Open media also reports about engagement of UAVs using manned aircraft. In September 2009, a manned US aircraft was launched to pro-actively kill a Reaper UAV over Afghanistan when its ground control station lost control over the drone[47]. There is a report of Israeli manned aircraft shooting down a UAV by using an Air to Air Missile (AAM) on 06 October 2012. The UAV was alleged to be an Iranian UAV, launched from Lebanon and tasked to carry out reconnaissance deep inside Israeli airspace[48]. Iran probably used its SU-25 fighter jets to fire on a US Predator UAV in November 2012.[49]

A Skewed Duel. While the hard kill options for the UAV using SAM system or worse, the manned aircraft, may appear to be straight, it is obviously badly out of balance in terms of cost consideration between the arsenal being used and the target being addressed. Besides, such hard kill measures can address individual targets or at best, a few; what about Swarms?, NAVs?, Micro/Mini UAVs? For these threats, such options may neither be viable, nor effective.

The Optimal Kill Option. The most potent option to take on a UAV is the soft-kill option. This killed method is anchored on debilitating the real-time data link between the UAV and its Ground Control

---

47   http:/www.popsei.com/.../when-drones-go-wild-airfoce-shoots-them-down.

48   http://www.state.com/.../Israel-shoot-down-drone-that.may-have-been-made-by-Iran. Accessed on 06 Jan 13.

49   http://www.guadian.co.UK>Glen Green World on security and liberty. Accessed on 06 Jan 13.

Station. The UAV's ground-to-air or air-to-ground data link signal to its GCS can either be jammed (which in all probability may result in UAV abandoning its mission and returning to base) or be hacked and exploited, in which case, the UAV's control can be taken over from the adversary and made compliant to the defenders' commands.

The Iranian Claim. In the above context, Iran claimed in Dec 2011 that its Electronic Warfare unit downed a US RQ 170 Sentinel UAV that had violated Iranian airspace. The photographs of the downed UAV show the aircraft relatively un-damaged except for minor damage on its left wing. This ruled out the possibility of a crash/ engine/ navigational malfunction. Experts feel that either it was a cyber/ electronic warfare attack system (indicating a soft kill) that brought the system down or it was a glitch in the command and control system.[50]

Fig 43 : US RQ 170 Sentinel Brought Down by Iran

Understanding the Soft Kill. The soft killing of UAVs involves attacking its communication links. Normally, UAVs use a line of sight radio link in the Military C band 500-1000 MHz or satellite communication in the KU band between 10.95-14.5 GHz (uplink band to satellite is normally 13.75-14.5 GHz and the down-link from the satellite is normally 10.95-12.75 GHz). If the UAV's communications to its GCS are jammed then the GCS operator

---
50 Dave Mazumdar, "Iran Captured RQ 170: How Bad is the Damage?" Air force Times Dec 09, 2011 quoted in http://www.en.wikipedia.org/wiki. Accessed on Jul 12.

becomes blind and the UAV flies around till it crashes or is out of fuel. In order to block the safe return route of UAV to GCS, both the above communication links must be jammed[51]. There was a report in the open media in September 2009 from the Russian aircraft industry site, Aura Port that efforts are on to incorporate ELINT based UAV killing capability in Ground Based Short Range AD Weapon System. For this, the existing radars of the GBADWS are being revamped with ELINT stations along with opto-electronic sensors, thus making such weapons capable of detecting the EM signature of the UAV. The soft kill is achieved by jamming the communication data links of the UAVs. There were also reports of Britain developing Directed Energy (DE) weapon package applicable for use against UAVs. The kill energy was to be derived on the higher power microwave route. Specific details about the weapon system are not known.

Laser Kill. Laser based killing of UAV is already an established option. Boeings' mounted a 1 KW laser system on its established Avenger platform and showed its capability to shoot down a handful of small UAVs. In the next phase of this weapon system, it is intended to install a 10 Kilo Watt solid-state laser on a Higher Energy Laser Mobile Demonstrator to take on various threat representative targets which includes UAVs/ UCAVs.[52] M/S Raytheon demonstrated its capability of close-in-weapon-system at the Farnborough Air Show in July 2010, wherein, a 50 Kilo Watt solid-state laser had a UAV in flames. The manufacturers are developing a low-cost directed energy laser system to complement the close-in-kinetic energy systems[53]. The same is being mounted on the proven Phalanx Weapon System. This Laser Area Defence Weapon promises to provide an effective kill solution to short range threats like the rockets, missiles, UAVs and such other targets. The kill speed is the speed of light and the magazine is nearly 'unlimited'.

---

51 http:/www.howtokillUAV. Accessed on 06 Jan 13.

52 http://www.news.cnet.com>news>cuttingedge>boeing-trucks-ahead-with-8-wheeling-laser-weapon. Accessed on 06 Jan 2013.

53 http:/www.raytheon.com/newsroom/feature/stellantgroups/laser-area-defence-system. Accessed on 06 Jan 2013.

 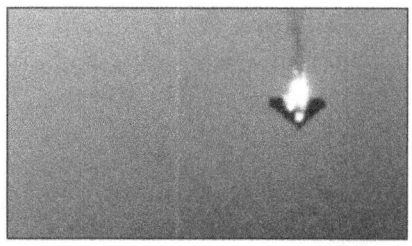

Fig 44 : Laser Avenger System   Fig 45 : LASER CIWS Shoots down a UAV

Kinetic Kill. Kinetic kill vehicles to take on UAV type of threats are also there in the field. Cougar is a kinetic kill solution manufactured by M/s QinetiQ of UK as a cost effective counter to low cost tactical UAVs. In this system, a low cost un-cooled long wave IR seeker is used to acquire the target and guide an interceptor on a collision course. The Cougar interceptor is designed to approach a UAV threat with closing speed that will be sufficient to cause catastrophic structural, damage to the target, giving an observable hard kill. Target UAV's low signatures in all wavebands, necessitates mid course guidance to place the Cougar interceptor in the collision course from where its on board seeker could acquire and home on to the target.

Fig 46 : Cougar Anti UAV System

Another Kinetic Kill Solution. Peregrine Eagle is another air launched decoy/drone system employed as an anti-UAV system in the Kinetic Energy kill domain having a range of 800 miles with a service ceiling of 40,000 feet. It employs high power microwave and is an electronic counter to UAV control.

Fig 47 : Peregrine Eagle : Anti UAV System

Man Over Machine? An Interesting Debate

Having seen both sides of the UAV/ UCAV, i.e. from the prosecution of the air threat, as well as, the defenders', a bit of crystal gazing into the future throws up some very interesting but difficult issues.

Face of the Future. As the future unfolds 4th Generation UAVs will come on the scene and will be more robotic than merely unmanned. They would be characterized by enhanced autonomy thereby removing the need for constant controlling (total vehicle autonomy). Artificial intelligence (AI) would take over more and more of decision-making. Automatically, returning capability would be further enhanced and so will be the craft's ability to assess situations and decide on course of actions by superior algorithms. Such machines will be more maneuverable and thus capable of air-to-air combat and multiple target strike operations with more precision and an ability to self-defend.

Joint Operations. Today, we are talking about unmanned platforms complimenting manned platforms in joint operations, wherein, the superior human intelligence intuition, grit, tolerance for ambiguity and instant decision making capability of the human intellect in cluttered situations is only to be complemented by the brute precision, range, stealth, immunity and versatility of the UAV/ UCAV platform. What about the opposite...? Difficult question ....

Man Over Machine or Other Way Round? When the autonomous machines laced with AI are able to make decisions, perform threat evaluation and threat prioritization based on probabilistic scenarios and generating machine options for optimal on board resource utility. Are we ready for unmanned platforms leading manned aircraft ? Are we ready for machines taking kill decisions unilaterally ? Are we ready for removing human override in the kill command decision by UAV/ UCAV platforms ? Are we ready for machine-over-man ? Probably not yet. Time and tide will unfold and resolve this rubic as the technology gallops forward and the terms of reference of this ethical but difficult debate get dynamically re-defined[54].

Conclusion And Some Take Aways

An Amazing Story. Thus goes the amazing growth story of the UAVs. Starting in their infancy way back in the mid nineteenth Century, coming of age through the two World Wars and growing wings of strength during the Cold War period and beyond. The employment of UAVs in operations right from the Vietnam War up to the GWOT clearly brings out how this versatile platform can handle a full spectrum of battle functions ranging from photo reconnaissance to jamming, leaflet dropping ESM (ELINT/SIGINT), creating chaff corridors, confuse enemy radars, act as decoys to draw enemy missiles to fire on infructuous targets, pre-ingress route reconnaissance, chemical agent detection, battle damage assessment, direct naval gun fire, ISTAR and finally, the hunter killer missions. All the crucial criticism of collaterals notwithstanding, the drone strikes in various parts of the globe, especially in the restive North West Frontier Province of Pakistan are going on unabated and are poised to increase in both quantum and lethality.

On Versatility. This book has attempted to bring out the versatility of the UAV platform and its multiple combat virtues. It has rubbished the talk of UAVs/ UCAVs being considered as total replacement of the manned platform. Highlighting the virtues and multiplicative

---

54 Presentation by Gp Capt Manoj Kumar, VSM Research Scholar at Centre for Air Power Studies during IMR Seminar 'Heli & UV India' on the topic 'A Defining Moment of the Debate;Manned Vs Unmanned Platforms. Accessed on 06 Jan 13.

advantages of joint operations, it has built the debate up to the present day dilemma of 'machine over man'.

Future of UAVs. This Chapter has also provided a glimpse into the future of the UAV bringing out the exciting morphing technologies, orinthropters, entomopters and more. Delving into the technological domain of mini/micro/nano UAVs, it has opened the gate of future technologies for the reader to see the face of tomorrow flagging the impact of nano and stealth in the UAVs.

The Flip Side. An attackers' delight is a defenders' nightmare has got amply illustrated in the text of the paper which has highlighted the multiple challenges in detecting the small, humble, low RCS, non-responsive, non-cooperative, day-night operational, stealthy and nano UAV.

Multiple Kill Options. All challenges notwithstanding, the paper has gone on to highlight the multiple kill options available to bring down the UAV and has also brought out the dilemma of the growing imbalance of sorts between the arsenal used (SAMs/ guns/ Kinetic Energy systems/ manned aircrafts) vis-à-vis the vehicle destroyed. Building on this cost imbalance, the paper has highlighted the advantages of soft kill (ELINT KILL, HPM based kill and Laser kill) of UAVs.

In the end, this Chapter in a bit of crystal gazing has thrown light on where do we reach on this growth path? It asks certain ethical questions on the eternal debate of manned vs unmanned and now machine-over-man. Are we ready? Only time will tell!

THE AMAZING GROWTH AND JOURNEY OF BALLISTIC MISSILE CAPABILITIES AND WHERE THE TECHNOLOGY IS LEADING TO

THE AMAZING JOURNEY OF THE GROWTH OF BALLISTIC MISSILE DEFENCE (BMD) CAPABILITIES ACROSS NATIONS AND WHERE THE TECHNOLOGY IS LEADING TO

"If you have the shield it is easier to use the sword[1]"

-Richard M. Nixon

"One sword keeps the other in sheath[2]"

Focus. This chapter chronicles the amazing journey of the BMD over the years and highlight show the current and likely future BMD capabilities across nations are flying high on the wings of cutting edge technologies.

The Era of Vengeance. In the context of putting up a defensive shield overhead against a potential missile attack, surely, the world has come a long way from the days of shock and total helplessness during World War II when Adolf Hitler's Vengeance Weapons, in the form of V2 rockets, came screaming down from the upper atmosphere, causing untold devastation and total havoc on a hapless population.

Fig 1 : 3000 + German V2 Rockets Targeted the City of London and Antwerp During WW II[3]

---

1 Wing Commander Anand Sharma, "Ballistic Missile Defence : Frontier of the 21st Century", KW Publishers Pvt ltd; 2010 p 129 accessed on 31 Aug 12

2 http:// www.brainyquote.com/quotes/keywords/sword_2.html accessed on 31 Aug 12

3 http://en.wikipedia.org/wiki/v2 accessed on 31 Aug 12

No Looking Back. The above V2 (Ballistic Missile) attack, which incidentally was the first in the world, not only sent the first human artifact in the outer space, but also, ended up causing huge casualties (7,250 allied forces and 12,000 forced labour killed in Germany while producing the weapons)[4]. The ballistic missile threat has never since looked back.

The Exponential Threat Growth. The said growth in the ballistic missile threat has been truly exponential in terms of range, reach (implying altitude), lethality and accuracy. The ballistic missile arsenal has grown through the TBM (range<300 km), to SRBM (range up to 1000 km), to MRBMs (1000-3000 km), to IRBM (range 3000-5500 km) and finally reaching the pedestal of ICBMs having intercontinental reaches (range>5500 km)[5]. Today, the names like SCUDS, Dong-Feng II, Ghaznavi (SRBMs) or Agni II, Taepo-Dong1, SS-3, DF-21 (MRBMs) or Agni IV, Shaheen III, DF-31, Shahab (IRBMs) or the mighty Galosh, Peacekeeper, A-135, SS-18 etc (ICBMs) sound familiar bells in the professional circles, assentinels of national security and pride for the countries which own these ballistic missiles[6][7][8][9]. So much so, that there is no target too far for the lethal punch and long reach of ballistic missiles, spanning continents and beyond[10].

Fig 2: SCUD   Fig 3: DF-21   Fig 4: SHAHEEN III Fig 5: PEACEKEEPER
(USSR) SRBM (CHINA) MRBM (PAK) IRBM (US) ICBM

---

4   Kenedy, Gregory P. "Vengeance Weapon 2, The V2 Guided Missile" Washington DC : Smithsonian Institution Press, 1983, accessed on 04 Sep 12

5   1 ibid p 5

6   www.en.wikipedia.org/wiki/short-range-ballistic-missile accessed on 04 Sep 12

7   www.en.wikipedia.org/wiki/medium-range-ballistic-missile accessed on 04 Sep 12

8   www.en.wikipedia.org/wiki/intermediate-range-ballistic-missiles accessed on 04 Sep 12

9   www.en-wikipedia.org/wiki/intercontentinential-missile accessed on 04 Sep 12

10 Andrew Erickson and Gobe Collins, "China's Ballistic Missiles : A Force to be Reckoned With", The Wall Street Journal, Aug 24, 2012, accessed on 04 Sep 12

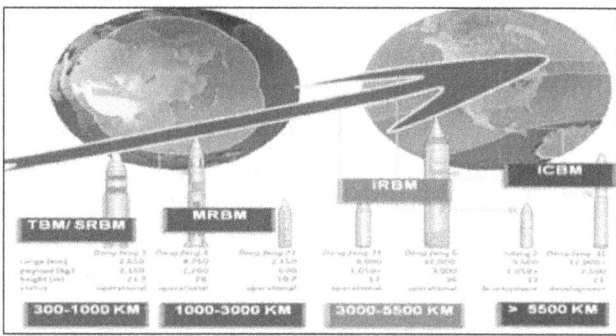

Fig 6 : For Ballistic Missiles, There is No Target Too Far

The Eternal Cause-Effect Relationship. To this ever sharpening 'sword' of ballistic missile threat, there is an ever strengthening shield of the BMD, coupled together in an eternal cause-effect relationship. This duel has spanned six decades plus, in a one-on-one adversarial relationship.

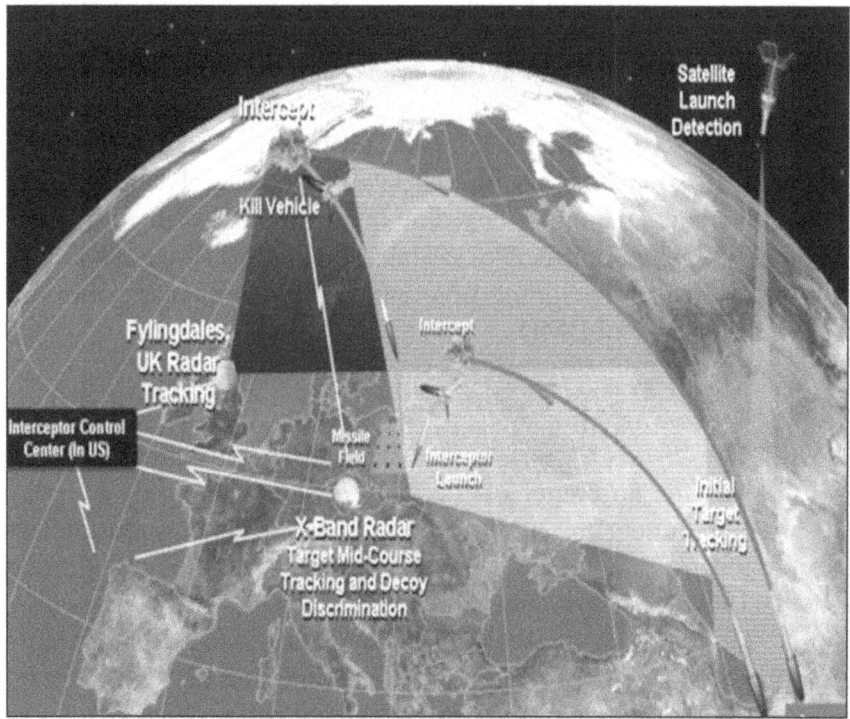

Fig 7 : The Eternal Cause Effect Relationship

The 50s and 60s. During the fifties and sixties, the Anti Ballistic Missile development by the two Superpowers in the cold war era was to go in for nuclear tipped interceptors carrying megatons of warhead. The strategy was to explode the nuclear warhead in the vicinity of the incoming ballistic missile thus devastating the threat. In this context, the mighty Spartan, Nike, Sprint, Sentinel of the US stood in defiance against similar ICBMs of USSR (Griffon, Galosh, Gazelle, Gorgon). The nuclear arms race and nuclear warhead based BMD had come of age.[11]

Fig 8 : Spartan BMD Missile (US)   Fig 9 : Gazelle BMD Missile (USSR)

The Realisation of Impracticality. By around the mid sixties, it dawned upon the then two Superpowers, that exploding nuclear warhead over each other's territory in the name BMD was both impractical, as well as, devastating, owing to the likely catastrophic outcome of many human casualties and unimaginable sufferings besides the gigantic effects of collateral damage. This gave rise to the felt need for developing non-nuclear hit-to-kill interceptors to support the cause of BMD.

HOE

   (a) **The Seed Experiment.** The above conceptual thought was the harbinger of the truly amazing 'Homing Overlay Experiment' in US; a programme to research non-nuclear means to intercept and kill ballistic missiles. Beginning 1960, the programme aimed to develop hit-to-kill technology based on homing sensors which would guide the interceptors onto the path of incoming ballistic missiles and kill them by the sheer force of their impact on collision.

---

11 1 ibid p 67-68 and 97 accessed on 04 Sep 12

(b) **The Follow-Up**. In a matter of eight years from 1976 to 1984, M/S Lockheed Martin, using the surplus Minuteman launch stages, did put out one exo-atmospheric homing hit-to-kill vehicle, having a long wavelength, long range IR seeker, targeted through a 'Direct and Homing Propulsion Section'.The innovative interceptor had 36 aluminium ribs with stainless steel fragments, to be deployed shortly after the intercept thus physically expanding the interceptor size. Put through four experiments (HOE 1 to 4 in Feb 1983, May 1983, Dec 1983 and Jun 1984 respectively), this technology demonstrator ended up destroying a Minutemanre-entry vehicle (RV) at a closing speed of 6.1 km/second and at an altitude of 160 km. The 'hit-to-kill-interception' had arrived.[12][13]

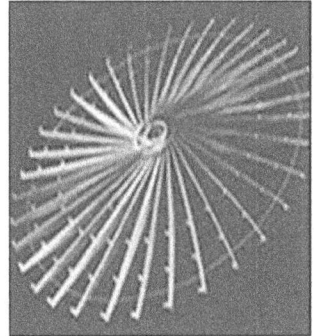
Fig 10 : The HOE

Fig 11 : The HOE Interceptor in the End Game

The Dawn of the Second Era

(a) **The Idea of SDI.** The second era of BMD in the field of building non-nuclear hit-to-kill interceptors dawned when on 23 Mar 83, President Ronald Regan of US proposed Strategic Defence Initiative or SDI for short. This ambitious project, riding on the wings of their cutting edge and futuristic technologies like High Power Lasers, Orbiting X Ray Lasers, Particle Beam Weapons, Brilliant Weapons etc, aimed to use ground and space based systems to protect US from strategic nuclear ballistic missiles of the adversary[14].

---

12 www.globalsecurity.org/space/systems/hue.html accessed on 04 Sep 12

13 www.airandspace.si.edn/collections/artifact.cfm?id=A39860223000 accessed on 04 Sep 12

14 www.nuclear files.org/menu/key-issues/missile-defense/history/reagan-on-stratagic-defense-intitiative.htm accessed on 04 Sep 12

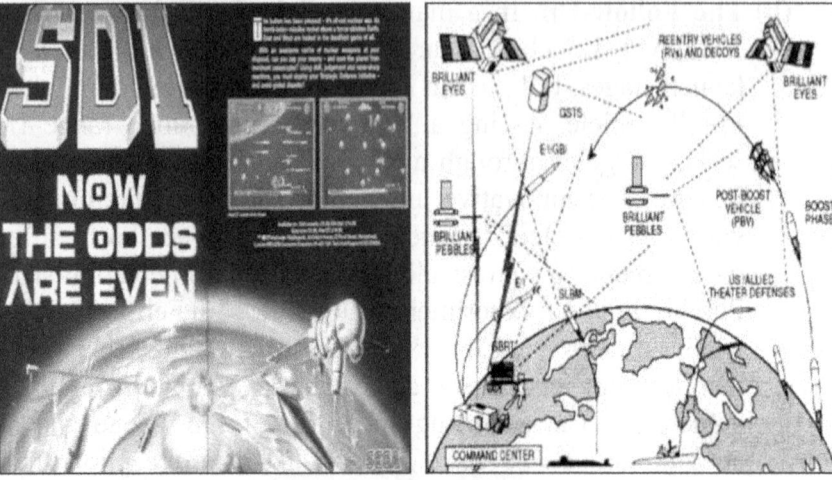

Fig 12 : The SDI - The Big Dreams

(b) **SDI Under Fire.** The critics dumped the SDI as unrealistic and unscientific. It was alleged that the SDI threatened to de-stabilise the strategic offensive doctrine of Mutually Assured Destruction (MAD), alongwith its abiding deterrant value, and would re-ignite an offensive arms race[15]. The press derided the SDI as 'Star Wars'. The American Physical Society concluded that the SDI was not only impossible with the existing technology, but also, more than ten years of research was needed to learn whether it might ever be feasible[16].

(c) **The Fate of SDI.** Faced with all the criticism, the SDI was re-christened as Ballistic Missile Defence Organisation (BMDO) and its emphasis shifted from National to Theatre missile defence and scope reduced from global to regional coverage. Further on, in 2002 the BMDO was renamed as Missile Defence Agency (MDA). The same continues till date and the Agency is central to all the BMD efforts of USA.

---

15 Cathal J.Nolan, "**The Greenwood Encyclopedia of International Relations**", S-Z. SDI, p-1600 accessed on 04 Sep 12

16 Mark Hertsgaard, "**Star Wars Works! 1996/06/07**", SALON quoted in www.en.wikepidiaorg/wiki/strategic-defence-initiative accessed on 04 Sep 12.

New Requirements for BMD Capability. As time elapsed and the threat from ballistic missiles got re-vamped upwards, the emerging BMD needs necessitated a ballistic missile defence in depth. It was envisaged that in such a defence, successive defensive layers would work synergistically to provide multiple opportunities to engage the incoming ballistic missile threat. Each layer was meant to force the attacker to develop complex counter measures to survive through one layer and press forward the attack through the next layer. Each layer was related to a particular range and altitude band of threat it could handle, as also, its chosen kill zone (endo/exo). Pertinent to note, that all layers still targeted the terminal phase of the incoming threat missile, simply because the technology would take us only that far. Needless to say, that the robustness (read effectiveness) of each layer was directly related to the independence (redundancy) built into it, which meant costs. The nations of the world, leading the BMD capability, accordingly went in for a tiered BMD architecture; the lower tier, upper tier and the strategic tier.

Lower Tier Systems. Designed for the defence of specific locations or targets, such systems were pitched against the adversary's SRBMs/ MRBMS with the end game in the endo-atmospheric region. A brief panorama of some significant Lower Tier Systems across the world is as under:-

(a) **The Patriot**.[17] [18] [19] Starting in 1981, the basic Patriot missile (MIM 104A) has seen many an upgrades. **PAC-1** (Patriot Advanced Capability-1) was designed with a lifted radar search angle (890), thus optimising it for a Ballistic Missile trajectory. The missile was upgraded to discriminate between artillery rockets and ballistic missiles and was capable of destroying, besides Ballistic Missiles, Electronic Counter Measure aircraft at stand off ranges. **PAC-2** featured further upgrades of radar search algorithms, besides war head revamping and an inserted delay between salvos to enhance kill effect. PAC-2 got further upgraded as **Guidance Enhanced Missile** (GEM), which had two versions, one especially optimized for Ballistic Missile

---
17 www.en.wikipedia.org/wiki/MIM-104-Patriot accessed on 04 Sep 12
18 www.mda.msl/system/pac-3.html accessed on 05 Sep 12
19 www.defense.gov/specials/missile defense/tmd-pac3.html accessed on 05 Sep 12

threat (PAC-2 GEM/T) and another optimized for cruise missile threat (PAC-2 GEM/C). Each of these missiles had a new faster warhead with delay reduced proximity fuze and a low noise seeker suitable for low Radar Cross Section (RCS) and stealthy targets. PAC-3 upgrade featured a totally new missile with a miniaturized warhead, altitude control motors for fine alignment of trajectory required for a precision hit-to-kill capability and a Ka band active seeker allowing the missile to drop its uplink and acquire the target on its own. The active radar gave the warhead a precision hit-to-kill capability that completely eliminated the need for proximity-fuzed warhead. Finally, PAC-3/ MSE (Missile Segment Enhancement) upgrade featured a new fin design and a more powerful rocket engine. This reportedly increased the operational capability of a typical PAC-3 missile by 50%. The ongoing PDB (Post Deployment Build) upgrade will increase the capability of the missile to discriminate between an Anti Radiation Missile, helicopter, UAV and cruise missile. There is also a talk to develop air-launched variant of PAC-3 for use in F-15 (C) Eagle, F-22 Raptor and P-8A Poseidon aircrafts.

Fig 13 : Patriot - Lower Tier System

(b) **MEADS (Medium Extended Air Defence System)**.[20] [21] [22] MEADS is another lower tier system which is being looked at as success or of Patriot (US), HAWK (Germany) and Nike Hercules (Italy). The co-operative venture between

---

20 www.en.wikipedia.org/wiki/Medium-Extended-Air-Defence-System accessed on 05 Sep 12

21 www.army-technology.com/projects/meads/accessed on 05 Sep 12

22 www.meads-amd.com accessed on 05 Sep 12

M/S Lockheed Martin (US), EADS (Germany) and MBDA (Italy) is claimed to be a 'next generation' point defence BMD solution. This system-of-system has facilities of open netted distributed architecture featuring a plug-and-fight capability allowing flexible-tailoring of multiple fire units. Air transportable in C-130, the base interceptor is PAC-3/MSE which has a forward-looking active radio frequency Ka band millimetric wave seeker providing precision hit-to-kill capability in the terminal phase. The basic configuration, which is due for initial deployment in 2014, features a surveillance radar, a multifunction tracking radar, light weight launcher and a PAC-3/MSE interceptor.

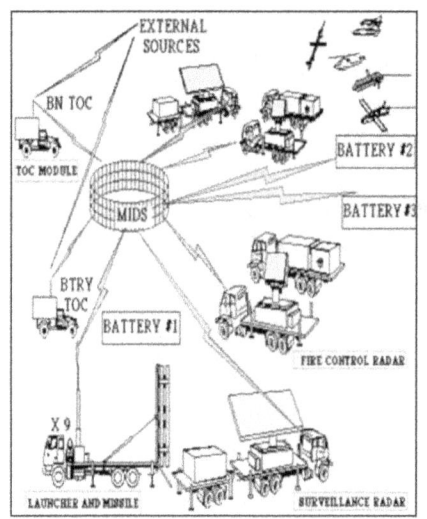 

Fig 14 : MEADS - Lower Tier System

(c) **Aegis BMD System.** In the field of sea-based BMD, this is another US lower tier system with a capability of post boost (mid-course) engagement of incoming threat. The interceptor is a Raytheon-RIM 161 Standard Missile in two versions, namely, SM2 and SM3 Light Weight Exo-Atmospheric Projectile (LEAP). As of Nov 2010, the System is already on board Five Ticonderoga class cruisers and 16 Arleigh Bruke class destroyers. Induction of land based version of Aegis Ballistic Missile System is also planned. This is called Aegis Ashore. The deployments are scheduled in 2015 (Romania) and 2018 (Turkey).[23] [24] [25]

Fig 15 : Aegis - Lower Tier System

(d) **Arrow (Hertz)**. Arrow Ballistic Missile System, a joint venture between the US (Boeing) and Israel (IAI), is also a lower tier theatre ballistic missile defence system which is claimed to be more effective than the Patriot system. It features a hypersonic (Mach 9) interceptor with an operational kill range of 90 km along with a directed high energy explosive fragmentation warhead.[26] [27]

Fig 16 : Arrow (Hertz) - Ballistic Missile System

---

23 www.en.wikipedia.org/wiki/Aegis-Baliistic-Defence-Systemaccessed on 07 Sep 12

24 www.mda.mil/system/aegis-bmd.htmlaccessed on 07 Sep 12

25 www.prnewswire.com/.../lockheed-martins-aegis-ballistic-missile-defence   accessed on 07 Sep 12

26 www.en.wikipedia.org/wiki/Arrow-(missile)accessed on 07 Sep 12

27 Bar Joseph, "**Israel's National Security Towards the 21 Century**", Frank Cass Publishers 2001 pp-153-154 accessed on 07 Sep 12

(e) **S-300 PMU Series of System.** Erstwhile USSR and now Russia, has a strong presence in the lower tier systems with several operational versions of S-300 SAM (NATO SA-10). This SAM system actually started out as an anti aircraft/anti cruise missile system. Subsequent variations (S-300 PMU) were optimised for ballistic missile defence role. The basic S-300 PMU-1had an operational range of 150 km, while its upgrade S-300 PMU-2 had an operational range of 200 km; both being third generation BMD systems. S-300 F (SA-N-6) was configured as a ship-based BMD system for terminal phase end game, while S-300 V (Antey 300) was designed for ground forces. The latest third generation ballistic missile system is S-400 Triumf (NATO SA-21 Growler) with an engagement range of upto 400 km[28] [29].

Fig 17 : S-300 PMU-1

Fig 18 : S-300 PMU-2

Fig 19 : S-300 PMU-V

Fig 20 : S-400 (TRIUMF)

Upper Tier Systems. Upper tier ballistic missile defence systems aim to carry out interception high in the atmosphere. Such systems are basically meant to take on the incoming ballistic missile of ranges upto 3500 km, this capability can however increase right upto the

---

28 www.sinodefence.com/army/surface to air missile/s300.asp accessed on 07 Sep 12
29 www.en.wikipedia.org/wiki/s-300_misileaccessed on 07 Sep 12

ICBM range of 10,000 km[30]. There-entry speeds may vary from 5-7 km/second. Terminal High Altitude Area Defence (THAAD) of US is one such system. It has an operational range in excess of 200 km and missile speed of Mach 8.24. The system is reported to have the capability of destroying the threat missiles as high as 150 km up in the atmosphere. System-to-system, it is capable of defending a surface area ten times that of Patriot. Currently served by the world's largest ground/air transportable radar (X Band Radar), its kinetic kill interceptor has the direct and altitude control system for precision terminal man oeuvre towards the target. Initial deployment of the system was slated in 2012[31][32]. The current status of this deployment has not yet been reported in the open-source domain.

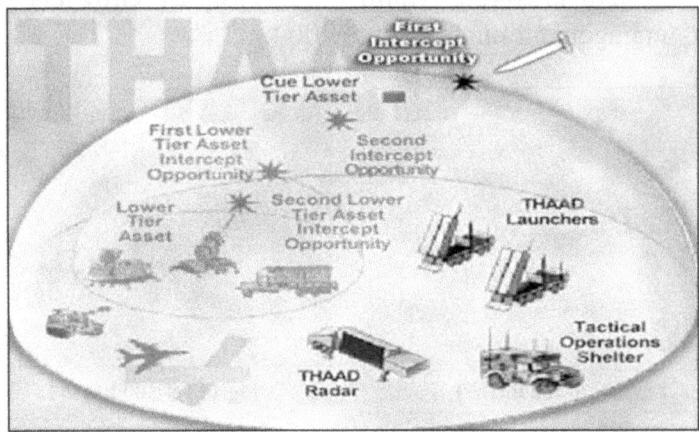

Fig 21 : THAAD - Upper Tier System

---

30 Ibid p 115

31 www. Military-today.com > Missiles accessed on 07 Sep 12

32 http:// www.en.wikipedia.org/wiki/Terminal - High-Altitude-Area-Defence accessed on 07 Sep 12

Strategic BMD Systems. Such BMD systems provide cover at macro scale, protecting city centres/seat of Governments/Key Areas, primarily against the adversary's ICBM threat. For example, the A135 (NATO ABM-3) system is a formidable ABM system deployed around Moscow. It is a two-tiered system with the outer tier having four launch complexes and 32 Gorgon exo-atmospheric interceptors, each with one megaton warhead and the inner tier having four launch complexes and 68 Gazelle endo-atmospheric interceptors[33][34]. In the same class, the US system is the Ground Based and Mid Course Defence System, aimed to protect continental US against limited attacks of ICBMS. This system features silo based missiles with exo-atmospheric kill vehicles[35][36].

 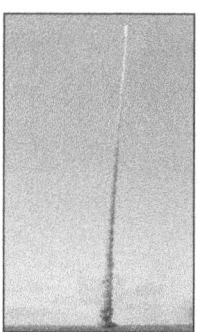  

Fig 22 : Pillbox Radar/ Russian A-135 System   Fig 23 : US Ground Based and Mid Course Defence System

WHERE THE TECHNOLOGY IS LEADING TO

How Far Have We Reached. With ever-moving wheels of time in the six decades gone by, since the V2 made its vengeance flights, time and technology has brought us to a pedestal in BMD capability where terminal phase interception in lower/upper and strategic tiers using hit-to-kill interceptors of various shades, shapes, sizes, ranges, reach and kill capability are a reality. The turning face of technology, however, never stops throwing up new challenges and corresponding trends to deal with such challenges. This portion of

---

33 http:// www.ausairpower.net/APA-Rus-accessed on 07 Sep 12

34 http:// www.en.wikipedia.org/wiki/A-135_anti_ballistic_misile_ systemaccessed on 07 Sep 12

35 http:// www.mda.mil/systems/ground.htmlaccessed on 07 Sep 12

36 http:// www.en.wikipedia.org/wiki/Ground _Based_ Mid_Course_ Defenceaccessed on 07 Sep 12

the paper aims to highlight such challenges and trends.

Fig 24 : Ever- Moving Wheels of Time   Fig 25 :Turning Face of Technology

Challenge 1: Boost Phase Interception (BPI).[37] [38] [39] The first challenge facing the BMD community is to take out the target missile in its boost phase of flight, i.e in the short period of a few fleeting minutes of its powered flight, immediately on launch.

Fig 26 : Challenges of BPI

(a) **Virtues.** The virtues of BPI are huge. Such an interception provides the best chance of a successful kill provided the capability exists. The exhaust of the missile is bright and hot. Its thermal signatures are really large providing the best case to the interceptor to track and kill the dumb, blind and bright payload with its fuel tanks still intact. The

---

37 http://www.far.org/spp/starwars/program/bpi.htm accessed on 07 Sep 12

38 http://www.ifpa.org/.../Ptalzgraff_BoostPhase_Missile_Defence_Capitolhill.htm accessed on 07 Sep 12

39 http://www.en.wikipedia.org/wiki/missile-defense accessed on 07 Sep 12

booster at this time is a large physical object providing a tangible RCS for detection and tracking. The target missile itself is very vulnerable, moving at a comparatively slow speed. With decoys/ counter measure not yet deployed the missile is one single object with no stages and no MIRVs. Early detection can provide opportunities for multiple engagements. Also, in the event of a successful intercept, the missile and its payload of weapons (which could well be nuclear/chemical/biological besides the conventional) will fall in the attacker's domain. A BPI intercept will reduce the 'safe havens' available to a hostile state. By 'safe havens' is meant the region of a country bounded by geographic and time constraints, from which a missile can be launched out of the range of the defender's BMD system.

(b) **Challenges.** While the virtues of BPI may be huge, it is not without enormous challenges. The very first is the surveillance challenge. In order to detect the target missile, thousands of kilometres away, a capability of a 'global look-see' is required. Also, mere detection is not adequate since tracking of such distant targets is a very complex task. This is so, because in the boost phase, the rate of change of acceleration mounts very rapidly. Besides this, the rocket plume effectively obscures the missile body making tracking a very great challenge. Not only the surveillance and tracking challenge is huge, even the interceptor has got to be very capable. Firstly, it must be placed close at hand so as to be able to be effective within the time frame at the boost location of the missile and secondly, it must achieve a precision kill on the speeding target missile, since hitting the booster will be of no gain.[40][41][42]

---

40 http://www.missile threat .com/repository/do clib/20030700-APSSG-bpi.pdf accessed on 07 Sep 12

41 http://www.global security.org>space>systems>BMD accessed on 07 Sep 12

42 http://www.aps.org/policy/report/studies/upload/boost-phase-intercept.PDF accessed on 07 Sep 12

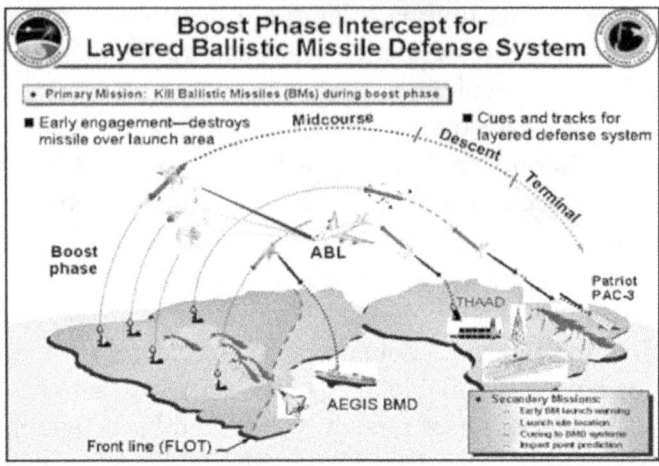

Fig 27-: Virtues and Challenges of Boost Phase Interception

Challenge 2 : Getting to the Next Best. If the Boost Phase intercept proves to be a far cry, given the current reach of technology and horizons of reality, the next best is to aim for intercepting the incoming ballistic missile in the post-boost phase ie, the ascent phase. This is a period immediately post the powered phase of the flight of the ballistic missile which could be as brief as 10 seconds or could be as long as tens of minutes. The virtues of such engagement still remain quite lucrative as the target missile is still on a predictable ballistic trajectory. In all probability, the counter measures, decoys and Multiple Independently Targeted Re-entry Vehicles (MIRVs) are not yet deployed. A total/partial kill at this stage will dilute the subsequent system requirements. Also a kill in this phase of the flight will have no debris effect falling in the target area. The challenges in carrying out this phase of engagement are almost as huge, (albeit a bit less) as the BPI. Since the thermal signatures of the target stand reduced, the detection and tracking of the target are that much more difficult and the end game has to be very precise since the warhead is now reduced to a small physical target. If the missile is carrying a 'Post Boost Vehicle (PBV)', it will cut both ways. In that, while the PBV will add to the RCS of the target ordnance, aiding in its detection and tracking, the likely deadly payload of PBV, i.e decoys, counter measures and penetration aids if realised before the interception event, will make the end game far more complex.[43] [44]

---

43 41 ibid
44 http://www.bits.de/NRANEU/BMD/MS Approach.pdf accessed on 09 Sep 12

Fig 28 : Complexities of Post-Boost Interception

The Challenges of BMD at Re-entry. By far, the challenges of carrying out interception at the re-entry stage far outreach the other phases. With multiple targets defined by MIRVs, decoys, countermeasures etc, the challenge is how to discriminate?, what to hit? and how to get at the warhead amidst the maze of decoys and countermeasures? The ionization plume of the melting debris/coatings at re-entry produces radar signatures much larger than the warhead vehicle.

The principle tracking challenge is the discrimination between the re-entry vehicle and debris/decoys/counter measures, travelling concurrently[45].

 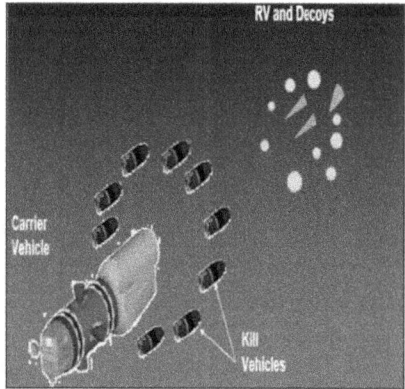

Fig 29 : Challenges of BMD at Re-entry

---

45 http://www.indian defense review.com/.../ballistic-missile-defence-system-for-India accessed on 09 Sep 12

Trends

General. In order to take on the challenges of BPI and post-boost interception, certain felt needs (trends) are clearly visible. These could be briefly recounted as under:-

(a) Acquiring the capability of a global-look-see.

(b) Synergizing sensor capabilities in multiple domains.

(c) Moving beyond the Kinetic Energy (KE) interceptors.

(d) Optimising battle management through a system-of-systems approach.

(e) Transiting from ASAT to BMD.

The Global Look-See

Trying to Meet the First Felt Need. As discussed, the first requirement for a BPI is the capability of the defender to achieve a capability of a global look-see, not only to detect missile launches from the, so-called 'safe havens' but also to guide 'appropriately based interceptors' onto them, before the boost phase runs out.

Technology's Answer. The technology's answer to a global look-see is, satellite based surveillance. This is based on the principle of detecting the radiation signatures of target missiles in boost/post boost phase. These signatures generally reside in the range of hundreds of kilowatts of infrared energy in the short wave (wavelength- 1.4-3 microns or µm, photon energy 0.4-0.9 electron volts or ev), medium wave (3-8 µm, 15-400 meV) and long wave (8-15 µm, 80-150 meV)[46]. These waves are detectable from great distances, even by the sensors placed on board the orbiting satellites.

Realities. Accordingly, the front runners in BMD (US, China, Russia) have operational capabilities in place providing satellite surveillance of likely ballistic missile threats, based on detection of infrared signatures of boosting missiles. These satellites are in Low Earth Orbits (LEO), Medium Earth Orbits (MEO), Highly

---
46 http://www.en.wikipedia.org/wiki/Infrared accessed on 09 S

Elliptical Orbits (HEO) or Geosynchronous and Geostationary Earth Orbits (GEO). Depending on their placement and the type of sensors on board, these satellites provide different formats of coverage varying in spatial spread and discrimination. Obvious to state, that with all other parameters remaining the same, the higher the orbit of the satellite, greater will be its spatial spread coverage and Lower will be its discrimination capability.

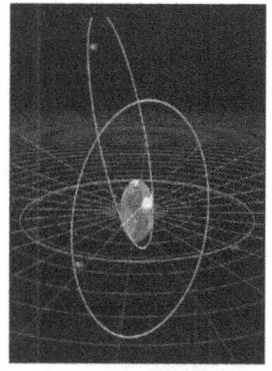

Fig 30 : Satellite Surveillance

Factual details of satellite based surveillance capability of some front line countries in BMD, are as under:-

(a) **US.** In US, the Defence Support Programme (DSP) (1991-2007 vintage is a system of 23 reconnaissance satellites in the geostationary orbits which forms the basic component of the satellite based early warning system[47]. DSP has been augmented by a satellite system under the generic name of Space Based Infrared Systems (SBIRS). This system is designed to meet US space surveillance needs for the first two to three decades of the 21st Century and will further enhance the capability of DSP. SBIRS (High) consists of four dedicated satellites (USA-205, 208, 209 launched in 2009 and USA 230 launched in 2011 respectively) in GEO and two host satellites (USA-184 and USA-200 launched in 2006 and 2008 respectively) in HEO. SBIRS-High is complemented by SBIRS-Low (renamed as Satellite Tracking and Surveillance System or STSS) which includes satellite in LEO. The primary purpose of SBIRS-Low (LEO) is to track ballistic missiles and discriminate between the warhead and other objects. SBIRS-Low was to have a system of 24 satellites but as per open sources information, the total SBIRS deployment as on date is six satellites with infrared sensors as indicated above. Other satellites in GEO under SBIRS are due to be launched in 2013. In essence, the US space based surveillance package is a multiple satellite capability built (and counting) over time; (DSP, SBIRS-

---

47 http://www.en.wikipedia.org/wiki/defense-support-programme accessed on09 Sep 12

High, SBIRS- Low or STSS).[48] [49] [50]

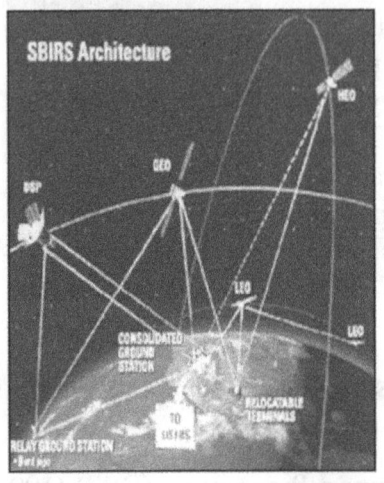

Fig 31: US SBIRS System Complex

(b) **China**. Building up its space-based surveillance and navigation is a key component of China's military modernisation. A large range of Yaogan/ Jianbing EO/SAR satellites are already in orbit; Haiyang ocean monitoring satellites, Tian Lian-1 relay satellites, Fenghuo 1 and 2 (Chinasat 22A) and Shentong-1 (Chinasat 20) military communication satellites[51]. It has been reported in the open media that three Chinese Yaogan Weixing-10 synthetic aperture and high resolution optical equipped satellites are flying in a formation akin to a maritime surveillance system[52].

---

48 http://www.defense industry daily.com/US/products/sbirs.html accessed on 09 Sep 12

49 http://www.sbirs.gov/ accessed on09 Sep 12

50 http://www.en.wikipedia.org/wiki/space-Based-Infrared-System accessed on 09 Sep 12

51 Andrew S. Erickson "Eye in the Sky", Proceedings, Apr 2010, PP. 36-41 quoted in Ian Easton, "**The Asia-Pacific's Emerging Missile Defense and Military Space Competition: Project 2049 Institute**", Dec 1 2010, p 42 accessed on 09 Sep 12

52 http://www.defence tech.org>Air Sea Battle accessed on09 Sep 12

Fig 32 : Chinese Satellite Constellation

(c) **ErstwhileUSSR/Russia.** Shrouded under the multi-discipline Kosmos series, the erstwhile USSR/present Russia, starting from 16 Mar 1962 till Jan 2010 has launched 2454 satellites of over 50 types. These include early warning satellites providing advance warning of missile attacks in the SBIRS mode, ELINT satellites and Ocean Surveillance satellites. All orbits are in use, viz, GEO/LEO/HEO etc[53]. Clearly, a capability in the class of the other two players (and more) as described above, exists.

Fig 33 : Russian Satellites in GEO/LEO/HEO

---

53 http://www.iap.era.int/sites/default/files/1.6.%20%/.20Mil%20%20SAAD.pdf accessed on 09 Sep 12

Synergizing Sensor Capabilities in Multiple Domains

What is the Trend? The second trend which is eminently visible towards achieving the global surveillance capability, is the augmentation of the sensor cover by synergizing the sensor inputs from the satellite based sensors with the inputs from long range ground / sea based radars.

Realities. The above trend is manifesting in all the leading BMD capable nations of the world. Some factual details are as under:-

(a) **US.** The long range (5556 km) 1995 vintage Precision Acquisition Vehicle Entry Phased Array Warning Systems or the PAVE-PAWS radar is being utilised to detect and track sea launched ballistic missiles (SLBMs) and ICBMs, besides the secondary mission of detecting and tracking earth orbiting satellites[54]. Another radar being used for mid-course coverage is the 1976 vintage COBRA DANE AESA radar. This huge (29 meter diameter) single face radar was basically erected to check the violation of SALT II Arms Limitation Treaty ex Russia. Yet another long range ground based radar which is a part of US GMBD is a sea-based X band Radar. This 2000-4700 km range radar is mounted on a fifth generation Norwegian design, Russian built twin-hulled semi-submersible drilling rig which can operate on high wind/high seas anywhere in the world.[55] [56] [57] The tremendous augmentation in sensor inputs which this radar can provide to the DSP-SBIRS-STSS combination, is self explanatory.

---

54 http://www.far.org/spp/military/programme/track/paws.htm accessed on 09 Sep 12

55 http://www.missile threat.com/missile defense systems/id.15/systems-detail.asp accessed on 09 Sep 12

56 http://www.en.wikipedia.org/wiki/Sea-based-X-Band-Radar accessed on 09 Sep 12

57 51 ibid p 40 accessed on 09 Sep 12

Fig 34: Cobra Dane Radar   Fig 35 : Mobile Sea Based X Band Radar

(b) **China.** Chinese are also using their long range (1000-4000 km) Over the Horizon - Backscatter (OTH-B) radars for augmenting the satellite based surveillance for missile warning. The overall surveillance effort is also supported by a fleet of six Yuanwang space tracking ships.

 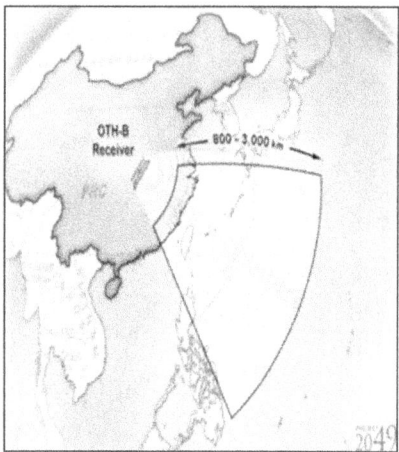

Fig 36 : Chinese OTH-B Radar Coverage

(c) **Russia.** Russia has nearly similar capabilities as that of US. Open sources report that in May 2012, Russians placed its new Voronezh-M long range missile warning radar on duty in Siberia. The radar has a key role in the missile warning chain of ground based radars. Another such radar is being built which will replace the Dnepr System; erstwhile Soviet Union's first generation phased array anti missile system.

With two Voronezh-M radars, an arc of 2400 (from India to USA) will stand covered. Compared to Dnepr (2500 km) system, this radar can detect ballistic missiles upto 6000 km. According to another open source, the land-based component of the early warning systems includes nine stations (called radio-technical nodes, ORTU). Each of them includes one or several radars (Dnepr, Daugava, Dnestr-space surveillance, Daryal U, Volga, Voronezh M etc). Five of the nine stations are located outside Russia[58].

Fig 37 : Voronezh-M Radar Station

Fig 38 : Dnepr Pulsed Radar Site

Fig 39 : Daryal Radar Site

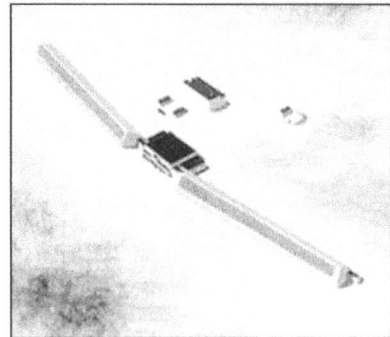

Fig 40 : Dnestr Radar Complex

---

58 http://www.en.rian.ru/military-news/21120123/173620853.html accessed on 09 Sep 12

Trend 3 : Moving Beyond the KE Interceptors[59][60][61]

Range-Time Deficit. When the challenge is to take out a boosting threat missile thousands of kilometers away in a matter of few fleeting minutes and seconds, the conventional KE interceptors fall short on many counts. These include, inevitable system reaction time from the decision of launch to the interceptor take off, the travel time of the interceptor to the boosting missile, (this in effect, is a function of the initial location of the interceptor and its speed of travel) and lastly, the kill capability of the interceptor booster warhead. Technically, if the interceptor is not placed sufficiently close to the launch location of the target missile (could even be in space, overhead the missile or at sea), system reaction and travel time make it impossible for the interceptor to reach the boosting missile before the boost phase runs out.

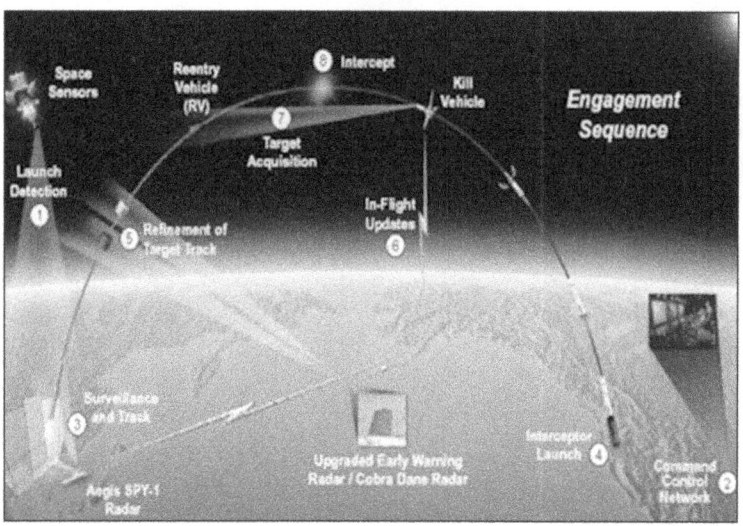

Fig 41 : Nuances of Range-Time Deficit during Interception

Trend. To address the above deficit, the emerging trend is to go for a new class of interceptors based on the directed energy of lasers or microwave or particle beams. The novelty of such interceptors are in two domains, viz, the 'kill mode' and the 'placement'.

---

59 http//:www.far.org/rlg/garwin-aps.html accessed on 09 Sep 12
60 http//:www.wlsweb.org/docs/SBLWP.pdf accessed on 09 Sep 12
61 http//:www.en.wikipedia.org/wiki/strategic-Defence.Initative accessed on 09 Sep 12

(a) **Kill Modes**. Each directed energy field is based on a specific kill mode. In the field of lasers, the coherent beams aim to concentrate hundreds of kilo joules of energy per square centimeter of the target delivering a shock impulse and causing thermal collapse of material leading to its destruction. In the case of High Power Microwave, the kill strategy is based on concentrating high power and high energy microwaves capable of carrying out adiabatic burn out of electronic components of target missile (called 'frying' of electrons') or detonation of electro explosive devices. In case of particle beams, the kill strategy is to produce an intense relativistic electron beam which is capable of providing a narrow band of RF output with power levels of several giga watts and output energies of hundreds of joules.

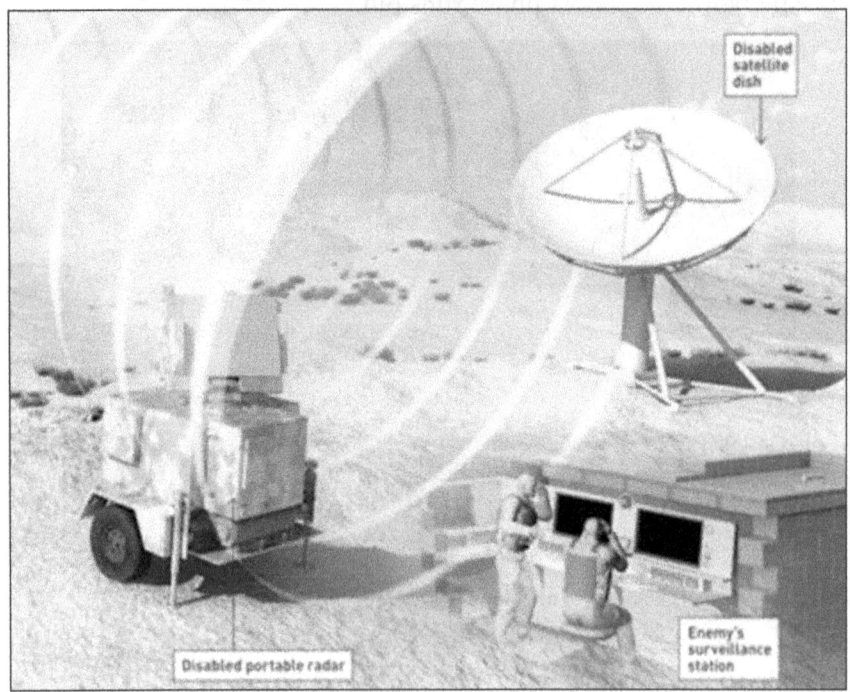

Fig 42 : Intense Relativistic Electron Beam

(i) **Placement.** The placement of such interceptor could be in any medium, viz land, sea, air, space. The only criteria is their meeting the reaction time deadline of impacting the target before boost phase runs out. It is pertinent to note, that two types of movements have to be taken into account. Firstly, the movement of the interceptor on its own steam (i.e its own boosters) to a location close enough to the missile so as to be able to launch its Directed Energy (DE) kill vehicle, and secondly, the very brief period of time (almost negligible) when the direct kill energy travels to the target and is incident upon the same. In an opportune case, if the interceptor is correctly placed (as will be desired) close to the target missile, so as to be in a position to fire the DE kill means ab-initio, then the time to target will be nearly nil.

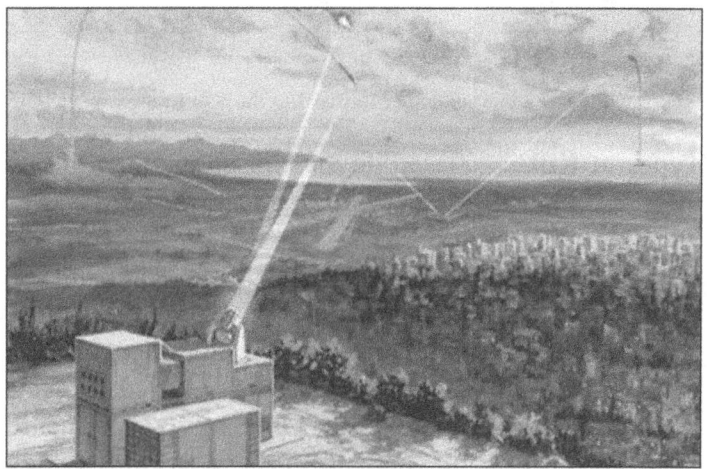

Fig 43 : Depiction of DE Weapons in Action

(ii) **Delivering the Kill Energy**. In DE interceptors, since the kill energy is being delivered at/near the speed of light, it amounts to 'shooting the bullet with the bullet'. For example, if a particle beam DE weapon is incident upon a re-entry vehicle travelling at 20,000 feet/second at an altitude of 50 km, the target will travel only five feet from shoot to strike.

Fig 44 : Delivering the Kill Energy

Types of Kill.[62] [63] Based on the chosen medium of DE kill means, there can be two types of kill, namely, electronic kill and thermal kill.

   (a) **Electronic Kill.** As stated, the electronic kill using the High Power Microwave (HPM) or Charged Particle Beam (CPB) media, aims to concentrate an impulse jolt of few hundred million electron volts (MeVs) carried through the KE of sub-atomic particles. Such energy, when incident upon the critical components of missiles/ re-entry vehicles, instantly alters the material composition and integrity of the material mass constituting the threat warhead/ RV, thereby destroying it in a catastrophic kill.

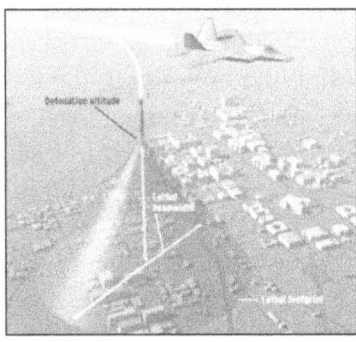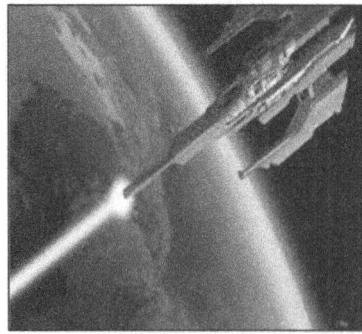

Fig 45 : Electronic Kill using HPM or CPB Weapons

---

62 http//:www.ausairpower.net/dew-ebsub.html accessed on 10 Sep 12

63 http//:www.acq.osd.mil/dsb/reports/ADA476320.pdf accessed on 10 Sep 12

(b) **Thermal Kill.** In the case of thermal kill using the lasing medium, the coherent beams instantly concentrate tremendous energy (hundreds of kilojoules as stated) delivering such a shock pulse that it leads to a thermal collapse of target material.

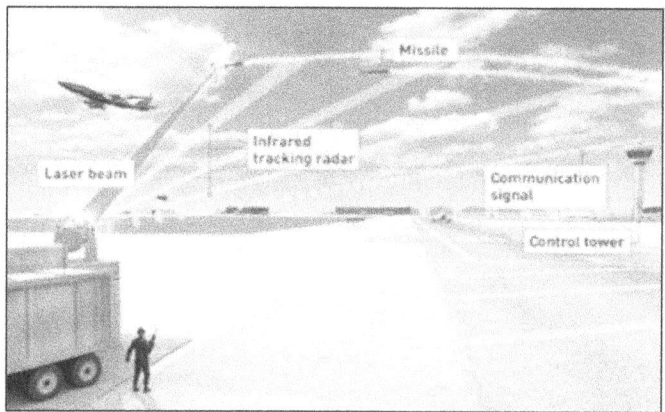

Fig 46: A Futuristic Thermal Kill Weapon

Challenges of DE Kill.[64] While the DE kill looks so fascinating making the reader believe the BPI is easily achievable, the same is ridden with huge challenges. Some details are as under:-

(a) **Placement Challenge.** One of the mega challenge is to ensure that the space based interception is at a typical altitude (say around 400 km) from where it can descend to its target missile in its boost phase, or alternatively, it could travel horizontally (instead of direct descent) along the missile trajectory with an aim to achieve the end game in post boost phase.

(b) **Battle of Wits.** To increase the odds against the defenders, while the attackers are constantly trying to reduce the booster burn out time to effectively shorten the BPI kill window, the defenders are trying to achieve higher and yet higher interceptor speeds (3000 km in 300 seconds!) in order to get to the target missile before the missile booster burn-out is

---

64 Robert L. Pfaltzgraff, Jr, "**Boost Phase Missile Defence : Present Challenges and Future Prospect**s", The Capital Hill Clubs 300 First Street, SE Apr 3, 2009 accessed on 11 Sep 12

still alive. The eternal cause- effect battle of wits is going on endlessly.

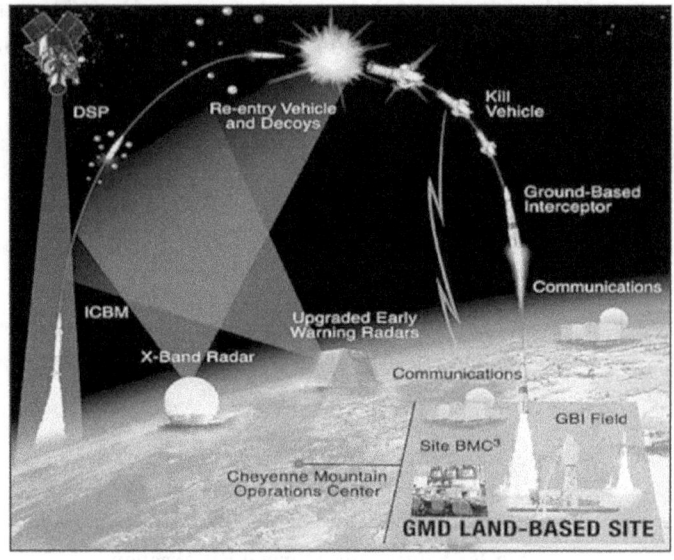

Fig 47 : The Eternal Cause Effect Battle

(c) **Challenge - IR Homing.** The IR homing interceptors are up against a huge challenge of keeping the look-out window super-cooled (against the intense frictional heat which the interceptor vehicle would pick up in its travel to target) in order to home-on to the target through a detectable temperature differential. Research is on to guide the interceptors to their boosting targets through commands sent from other satellites residing in cool-space backgrounds.

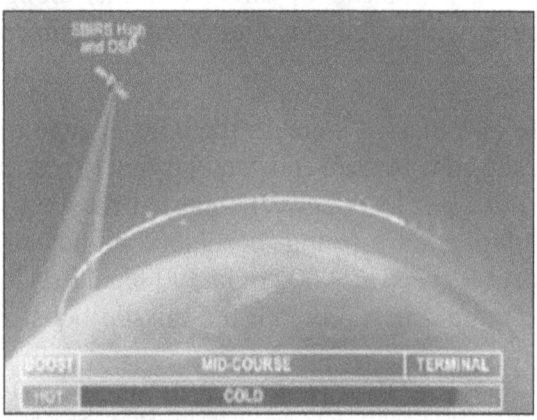

Fig 48 : Challenges of IR Homing

(d) **Other Challenge**. There is yet another challenge of orbiting huge interceptor loads in the space and the sheer quantum of satellites required to cover a particular threat domain. Also, there is a very big challenge of generating sufficient power to achieve the type of catastrophic electronic/ thermal kill as enumerated above. For example, in case of charged particle beam weapons, the principle challenge is to produce millions of joules of energy in nano-second pulses through series of cyclotrons/ synchrotrons etc. In case of laser kill the coherent beam must be adequately power-packed. Innovative technologies are being employed to enhance the power of laser beams, eg, the US High Power System Interception (HPSI) calls for successive firing of low power lasers followed by a high power kill laser in the end game thus reducing cumulative energy requirement over the entire engagement spectrum[65].

(e) **The US Airborne Laser Test Bed (ALTB) Experiment**. In Feb 2010 the above technology was demonstrated, wherein, a short range threat representative ballistic missile was launched from a mobile at-sea launch platform. Within seconds, the Airborne Laser Test Bed (ALTB) used on-board sensors to detect the boosting missiles and used a low-energy laser to track the target. The ALTB then fired a second low energy laser to measure and compensate for atmospheric disturbances. Finally the ALTB fired a mega-watt class High Energy laser heating the boosting missile to critical structural failure. The entire engagement occurred within two minutes of the target missiles launch, while its rocket motors were still thrusting[66]. Further research and experiments in this field are currently on and full operational status has not been reported anywhere in the open source information[67] [68]. Another research area is to have the power station on ground with the capability to reflect laser kill

---

65 http/www.mza.com/.../acc 2020/MZA_PS_ACTB_11/123pd... US accessed on 11 Sep 12

66 http/www.mda.mil/news/10 news 0002.html accessed on 11 Sep 12

67 http/www.mda.mil/news/10 news 0011.html accessed on 11 Sep 12

68 http/www.mda.mil/news/10 fyi 0001.html accessed on 11 Sep 12

energy through the science of adaptive optics. As a counter, one laser kill avoidance measure is to use mirror/adaptive optics to reflect/deflect a large part of incident energy or to rotate the boosting missile so that the lasing energy is absorbed by a larger surface area. Indeed one-on-one!

Fig 49 : The Airborne Laser Test Bed

Current Developments. The current pace of development suggests that directed energy interception could possibly become the centre piece of the futuristic layered BMD architecture. The fundamental advantage being of delivering the kill energy at/near the speed of light and its instantaneous arrival on the target. Another important research field is kill assessment. This is essential, keeping in mind that while a boost/post-boost DE intercept may kill a booster/missile (albeit partially), its capability to deploy MIRVs/decoys may still remain intact. Also, there is a need to check out whether the interceptor seeker could actually discriminate or not between the target warhead and the rocket plume boosters.

Optimising Battle Management

A Hypothetical Scenario. Just sample this. A boosting missile has just been launched thousands of kilometers away in adversary's land (almost half way across the globe). Within seconds, the same is required to be picked up by IR or other sensors based on orbiting satellites in space and information passed down in near real time.

Possibly, this information needs to be corroborated very quickly though other ground / sea-based radars. Almost in near real time, the decision is to be taken to command the interceptor through quick steps of interceptor selection (one on many) and accurate weapon designation. This is to follow with real time guidance / control of speeding interceptor till its active seekers on-board take over for a precise end game. Alongside this, go the challenges of accurate target selection, real time precise home-on and the capability to distinguish between the warhead and thrusting boosters on the warhead and much hotter, rocket plumes etc.

System-of-Systems. To battle-manage such complex engagement scenarios calls for a system-of-systems battle management architecture. Literally, a system-of-systems is a collection of task oriented or dedicated systems that pool their resources and capabilities together to create a new, more complex system which offers more functionality and performance than simply the sum of constituent systems[69].

System-of-Systems in the Parlence of BMD.

(a) **The Concept.** The system-of-system in the BMD parlence is based on the above concepts. It networks multiple inputs and exercises real time control in multiple systems and controls. In essence, it processes a variety of space based data being obtained from multiple satellites, fuzes them and sifts through them continuously to detect global threats with a special eye on threat-prioritized domains, continuously keeps augmenting space based surveillance data with ground/ sea based radars to obtain theatre wise comprehensive air picture. At the same time, the system keeps count of various types of interceptors in terms of their various parameters like range, reach, location, readiness status etc.

(b) **The Action.** In case of a threat detection a flurry of actions unfold at great speed like initial detection, positive identification and recognition through combination of surveillance inputs from various sources, interceptor situation, target designation and real-time control of interception

---

69 http/www.en.wikipedia.org/wiki/system-of-systems accessed on 11 Sep 12

till active seeker on board takes over for the end game. In essence therefore, the system-of-system will coordinate and network three major activities, viz, threat detection through satellite and ground/sea/air based surveillance and tracking resources, real time processing of surveillance inputs (through detection, identification, recognition, prioritization, interceptor selection and target designation) and interceptor launch and control of engagement.

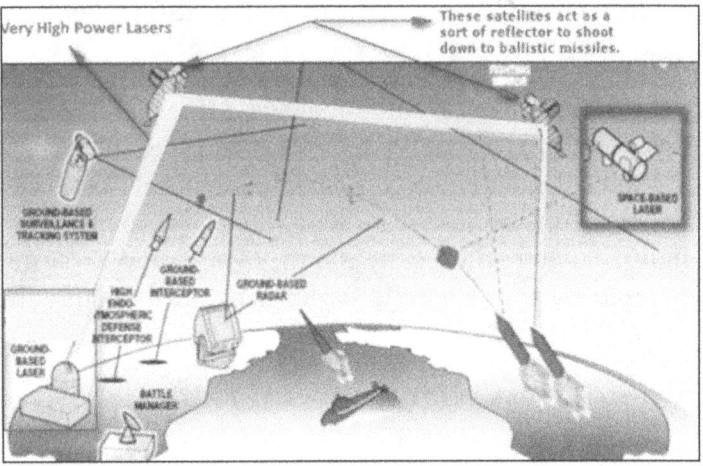

Fig 50 : System of Systems Approach

Reality. All the front-runners in the BMD race are putting up and perfecting such BMC2 architecture through system-of-systems approach, e.g, in US, the three network elements are DSP-SBIRS-STSS for satellite surveillance duly augmented by ground/sea based

Fig 51 : Joint Tactical Ground Station (JTAGS)

radars (PAVE-PAWS, COBRA DANE, X Band), Tactical Detection and Reporting Centre or TACDAR for processing a variety of space based data, fuzing it with other inputs and relaying the same across theatres and Joint Tactical Ground Station or JTAGS for further processing and disseminating missile information and alerting and cueing of interceptors to targets.

Transiting From ASAT to BMD

The Game Changer Events.[70] [71] The ongoing race for the space weaponisation/ militarisation amongst many a space-faring nations changed forever on 11 Jan 2007 when a Chinese direct Ascent Anti Satellite (ASAT) weapon (SC-19, based on a modified DF-21 ballistic missile) and carrying a kinetic kill vehicle or KKV (modified HQ 19 SAM) destroyed an aeging Chinese weather satellite (FYIC of the Fengyun series) in the polar orbit at an altitude of 865 Km with KKV travelling at 8 km/second. A year later on 11 Feb 2008 USA destroyed a malfunctioning US spy satellite USA-193, using a RIM 161 Standard Missile (SM-3) in a mid course interception 100 miles over the Pacific using direct ascent. The ASAT capability had come of age.

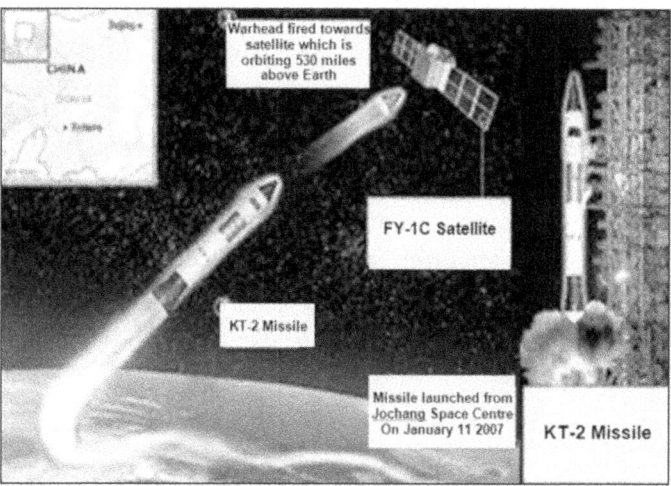

Fig 52 : Chinese ASAT Test

---

70 http//www.en.wikipedia.org/wiki/2007-chinese-anti-satellite-missile-test accessed on 12 Sep 12

71 http//www.far.org/sgf/crs/row/RS22652.pdf accessed on 12 Sep 12

Other Reported Chinese Endeavours. As per the open source information, China, prior to its Jan 11 2007 test, had conducted two or three previous such tests which had even tested a direct energy, laser ASAT Weapon on a US intelligence satellite. Subsequently, as a part of its manned Shenzhou-7 mission in 2008, China conducted a micro satellite test with implications for development of a co-orbital ASAT weapon[72]. Forging ahead with multiple options, China is developing directed energy microwave, particle beam and laser ASAT weapon capability. This capability could take on satellites in GEO and LEO under a soft kill[73].

Point at Issue-Co-Terminity. The growing ASAT capability of US and China has been quoted in context to flag the point at issue, i.e "If a ASAT weapon can take out a satellite in orbit why can't it take out a boosting missile in the post-boost or mid course phase" i.e there is a co-terminity between the ASAT capability and the BMD. According to open source information, post the success of ASAT experiment, China has tested an ASAT based mid-course interceptor thus marking a transit from the ASAT to BMD capability. This test is widely suspected as being the second successful test of Chinese direct ascent ASAT system[74]. It would be logical to grant similar transit capability to US which actually leads the World today in BMD technologies and has a direct ascent ASAT kill under its technological wings.

Still Baby Steps. While this paper has brought out the feasibility of the co-terminity of ASAT capability into the BMD domain, open sources yet do not report completion (implying operationalization) of such a transit, as such the space remains contested environment for placing the BMD capabilities[75]. Way back in 2002, Paul Wolfowitz, Deputy US Secretary of State for Defence confirmed the US Administration's ambition to see that weapon in space

---

72 Ian Easton, **"The Great Game in Space : China's Evolving Assault Weapon Programme and their Implications for Future US Strategy"** the Project 2049 institute occasional paper, Jun 24, 2009 accessed on 12 Sep 12.

73 Wendell Minnick **"China Missile Test Has Omnious Implications"**, Defense News Jan 09, 2010 accessed on 12 Sep 12.

74 73 ibid

75 **"High Frontier"** Volume 6 No 3 May 2010, pp-3 accessed on 12 Sep 12

become a part of the multi-tiered missile defence[76]. Domain experts feel that applying the ASAT capability to take on ballistic missiles through space based interceptions in dual use is a near certainty in the foreseeable future. Also, according to open source domain, the Chinese space programme has been from its very inception, a fundamental military endeavour which is pushing its lead rival down slippery slope of space weaponisation in the ASAT, as well as, BMD domain[77].

The End Piece. Thus goes the amazing growth story of the BMD. Born out of the helplessness, shock and awe of a hapless humanity being ruthlessly mauled by the vengeance of Adolf Hitler, initially basing its existence on megatons of nuclear warheads atop ICBMs waiting to collide against incoming ballistic missiles, dawning of the realization of impracticality of a nuclear catastrophe in the name of BMD, trail - blazing a new course with non-nuclear interceptors, gaining in strength through multiple layers over time and now extending its ever-growing wings into the limitless opportunities in space - the ultimate high ground. Possibly, the abiding battle between the sword and shield will go on and on and on, eternally.

---

76 Wing Commander Anand Sharma "**Ballistic Missile Defence, Frontier of the 21st Century**", KW publishers Pvt Ltd p-212 accessed on 12 Sep 12

77 Kevin Pollpeter, "**Building For the Future : China's Program in Space Technology and US Response**", Strategic Studies Institute 2008, p-5

Index

A

A135 (NATOABM-3) 51
ADM-20 Quail v, 5, 6, 77, 80
Aegis BMD System 47
AMAZING GROWTH and JOURNEY OF Unmanned aerial vehicles
  Evolutionary Path 4
Amazing World of Orinthropters vi, 21, 80
APACHE Armed Attack Helicopter 14
AQM 34 6
AQM- 34 v, 6, 80
Arrow (Hertz) 48
Arrow (Hertz) - Ballistic Missile System vii, 48, 82

B

Backscatter (OTH-B) radars 61
Baitullah Mehsud 10
Barak vi, 29, 30, 77, 81
Barak SRSAM 29
Black Widow vi, 21, 22, 77, 80
BMD 39, 41, 42, 43, 44, 45, 47, 49, 51, 52, 53, 55, 56, 57, 60, 70, 71, 72, 73, 74, 75
Boeing B-52 Stratofortress 5
Boeing X-45 vi, 17, 77, 80
Bringing Down The UAVs v, 80

C

Charged Particle Beam (CPB) 66
Chinese ASAT Test viii, 73, 83
Chinese OTH-B Radar viii, 61, 82
Cobra Dane Radar vii, 61, 82
Comanche v, 15, 80
COMANCHE UAV 14
Compass Copes 6
Cougar Anti UAV System vi, 33, 81

D

DARPA 14, 21
Daryal Radar Site viii, 62, 82
Dassault Aviation 17, 25
Dassaults Aviation nEUROn v, 17, 80
Defence Advanced Research Project Agency 14
Desert Hawk vi, 22, 77, 80
DF-21 vii, 40, 73, 81
DH 82B Queen Bee drone 5
Dnepr Pulsed Radar Site viii, 62, 82
Dnepr System 61
Drones 4, 10, 11

E

ELINT 6, 32, 35, 36
ESM Based Sensing 28
Exodrones 7

F

F-15 (C) Eagle 46
F-22 Raptor 46
F-22 Raptor, Contemporary Air Threat Air v, 3, 80
F-35 Lightening II v, 16, 80

G

Galosh 42
Gazell 42
Gazelle BMD Missile (USSR) vii, 42, 81
Geostationary Earth Orbits (GEO) 57
German multirole UAV LUNA 8
German V2 Rockets vii, 39, 81
Global War on Terror 10
Gorgon 42
Griffon 42
Ground Control Station 9
Guidance Enhanced Missile 45

H

HALE UAV 'Global Hawk' v, 9, 80
Highly Elliptical Orbits (HEO) 56
High Power Microwave (HPM) 66
HOE vii, 42, 43, 81
Humming Bird vi, 22, 80
Hunter killer missions 10

I

ISTAR mission 24

J

JAS-39 Grippen platform 17
Joint Tactical Ground Station (JTAGS) viii, 72, 82

K

Kettering Bug v, 5, 80
Kinetic Kill 34

L

Laser Avenger System vi, 33, 81
LASER CIWS vi, 33, 81
Laser Kill 32
Laser Range Finder 27
Lightening Bugs 6
Low Earth Orbits (LEO) 56
low radar cross section 26
Luna v, 8, 80

M

MEADS vii, 46, 47, 78, 82
MEADS (Medium Extended Air Defence System) 46
Medium Earth Orbits (MEO) 56
Mid 1930 - ' Queen Bee' Drone v, 5, 80
Mobile Sea Based X Band Radar vii, 61, 82
Morphing 19, 20, 36
Morphing Technology vi, 20, 80
MQ-8 Fire Scout 17
M/S Lockheed Martin 16
M/s QinetiQ of UK 33

N

nEUROn v, vi, 17, 25, 78, 80, 81
Nike 42
Northrop Grumans' Firescout Un-

manned Autonomous Helicopter vi, 18, 80
Northrop Gruman X-47 17

O

Operation Anaconda 9
Operation Desert Storm 7
Operation Enduring Freedom 8
Optimal Sensors 27
Orinthropters 20
OSA-AK vi, 29, 81

P

P-8A Poseidon aircrafts 46
Passive Sensing 27
Peregrine Eagle vi, 34, 78, 81
Peregrine Eagle : Anti UAV System vi, 34, 81
Perigrine Eagle : Anti UAV System 34
Phantom Ray UAV vi, 25, 81
Phased Array Radar Technology vi, 27, 81
Pioneer UAV v, 8, 80
Pointer UAV v, 8, 80
Predator v, 8, 9, 10, 11, 12, 13, 15, 17, 26, 30, 79, 80

R

Raytheon-RIM 161 Standard Missile 47
RC-135 9
Reaper 11
Reginald Denny with 'RP-1' v, 5, 80
Remotely 'Piloted' Vehicles (RPVs) 4
RQ4A Global Hawk 8
RQ 170 Sentinel vi, 25, 31, 81
Russian A-135 System vii, 51, 82
Ryan 147 Lightning Bug Series v, 80
Ryan Firebee v, 6, 78, 80

S

S-300 PMU-1 vii, 49, 82
S-300 PMU-2 vii, 49, 82
S-300 PMU-V vii, 49, 82
S-400 (TRIUMF) vii, 49, 82
Samson and Delilah 7
SCUD vii, 40, 81
Sentinel 42
SHAHEEN III vii, 40, 81
short take off and vertical landing 16
SIGINT 6, 7, 12, 35
SM2 Light Weight Exo-Atmospheric Projectile 47
SM3 Light Weight Exo-Atmospheric Projectile 47
Smart Warfighting Array of Reconfigurable Modules 22
Spartan 42
Spartan BMD Missile (US) vii, 42, 81
Sprint 42
Spyder vi, 30, 81
SRSAM Domain 29
Star Streak vi, 29, 81
Strategic BMD Systems 50
Strela 10M vi, 29, 81
SU-25 Frog foot Warplane 13
SWARM 22
SWARMs of Quadrotors vi, 23, 81
SWAVNET vi, 25, 81

Switchblade Suicide Drone vi, 22, 80

T

Teledyne Ryan 124 R RPVs 6
Teledyne Ryan 1124 v, 7, 80
THAAD vii, 50, 82
Thermal Kill 67
Threat Metamorphosis 3
Tier II Predator UAV 8

U

UAVs 3, 4, 5, 6, 7, 8, 9, 10, 11, 12, 13, 15, 17, 19, 20, 21, 22, 23, 24, 25, 26, 27, 28, 29, 30, 31, 32, 33, 34, 35, 36

UAV X-47B 15
Use of Balloons as Aerial Threat Vehicle v, 4, 80
US RQ 170 Sentinel vi, 31, 81
US SBIRS System Complex vii, 58, 82

V

V2 rockets 39
Vintage Air Threat v, 3, 80
Voronezh-M long range missile warning radar 61
Voronezh-M Radar Station viii, 62, 82

W

War on Terror 11

X

X 47-B v, 16, 80

Y

Yaogan/ Jianbing EO/SAR satellites 58

Z

Zhuhai Air Show 2010 11

www.ingramcontent.com/pod-product-compliance
Lightning Source LLC
Chambersburg PA
CBHW060849050426
42453CB00008B/916